Lecture Notes in Computer Science 758

Edited by G. Goos and J. Hartmanis

Advisory Board: W. Brauer   D. Gries   J. Stoer

Monique Teillaud

Towards Dynamic Randomized Algorithms in Computational Geometry

Springer-Verlag

Berlin Heidelberg New York
London Paris Tokyo
Hong Kong Barcelona
Budapest

Series Editors

Gerhard Goos
Universität Karlsruhe
Postfach 69 80
Vincenz-Priessnitz-Straße 1
D-76131 Karlsruhe, Germany

Juris Hartmanis
Cornell University
Department of Computer Science
4130 Upson Hall
Ithaca, NY 14853, USA

Author

Monique Teillaud
Institut National de Recherche en Informatique et en Automatique
B. P. 93, F-06902 Sophia Antipolis Cedex, France

CR Subject Classification (1991): I.3.5, F.2.2, G.2

1991 Mathematics Subject Classification: 68U05, 68Q20, 68Q25, 52B55

ISBN 3-540-57503-0 Springer-Verlag Berlin Heidelberg New York
ISBN 0-387-57503-0 Springer-Verlag New York Berlin Heidelberg

Typesetting: Camera-ready by author
Printing and binding: Druckhaus Beltz, Hemsbach/Bergstr.
45/3140-543210 - Printed on acid-free paper

Preface

This book is based on my *Thèse de Doctorat* defended on December 10, 1991, at *Université de Paris-Sud, Orsay, France*.

I am grateful to Jean-Daniel Boissonnat who was an active and dynamic advisor. He animates Prisme, the team I am in, at I.N.R.I.A.–Sophia-Antipolis.

Je remercie Bernard Chazelle and *ich danke* Kurt Mehlhorn, who both honored me by accepting to be external referees. I thank the members of the thesis committee, who encouraged me to publish these Lecture Notes, for their interest in my work : Dominique Gouyou-Beauchamps (President), Jean Berstel, Jean-Daniel Boissonnat, Claude Puech, and Jean-Marc Steyaert who, in addition, wrote a referee report and made several useful remarks to me.

The core of this thesis corresponds to several presentations in conferences, and articles in international journals, co-authored with Jean-Daniel Boissonnat, Olivier Devillers, Mariette Yvinec, Stefan Meiser (Max Planck Institut für Informatik, Saarbrücken, Germany) and René Schott (C.R.I.N., Nancy, France), depending on the chapters. I want to thank them all, as well as André Cérézo and Claire Kenyon, for the discussions we have had.

If this book happens to be famous in the future, it will undoubtedly be due to Bernhard Geiger, who made all the pictures at the beginning of the chapters, showing that research has several funny aspects !

I omit here to mention all the people I thanked in my thesis, though I do not forget them...

September 1993   Monique Teillaud

This work was supported in part by the ESPRIT Basic Research Actions 3075 (ALCOM: ALgorithms and COMplexity) and 7141 (ALCOM II).

The pictures were drawn using the peerless interactive drawing preparation system Jtdraw .

Table of Contents

Introduction

Computational Geometry concerns itself with designing and analyzing algorithms for solving geometric problems. It is a recent field of theoretical computer science, that has developed rapidly since it first appeared in M.I. Shamos' thesis [Sha78] in 1978.

The field has already reached a high level of sophistication, sometimes being too theoretical to be really useful: to achieve optimal running times, it is occasionally necessary to design very complicated algorithms, using data structures whose implementation is difficult.

Though the importance of such research is unquestionable, it can be useful, from time to time, to develop more practical algorithms. This is all the more true when their development also yields attractive theoretical problems while using rigorous methods. One such method is the use of randomized algorithms, introduced to the field by K.L. Clarkson in 1985 [Cla85]. Once introduced, these algorithms became more and more popular, turning into one of the hottest areas of recent years (sometimes too hot to my taste, though racing is stimulating!).

Randomization avoids the use of complicated data structures, and yields efficient algorithms, both from the point of view of theoretical complexity as well as real-life usage. It must, however, be noted that, in the literature, although emphasis is often put on practical efficiency, it is usually only mentioned and seldom experimentally tested.

The running times of these randomized algorithms will only be efficient in the average case. More precisely, the execution of the algorithms has a random aspect, but the algorithms themselves remain purely deterministic, and give an unique and exact output. Moreover, we emphasize that their analysis assumes no special configuration of data, while classical average-case analysis usually assumes some probabilistic distribution on data.

Randomized analysis averages over all possible executions of the algorithm, but assumes worst-case data. Consequently, it is a realistic model of analysis: in practice, interesting data are often points measured on objects, their distribution is far from being homogeneous, and they do not obey any probability assumption in space. Randomization is interesting from this point of view if the expected complexity of a randomized algorithm is better than the worst-case complexity of the best deterministic algorithms known for the same problem, or if equal, when the algorithms are simpler (which is very often the case).

Introducing some quantities representing the complexity of the outputs in the analysis is also possible, and leads to algorithms that are output-sensitive to a certain extent.

We were particularly interested in designing dynamic algorithms, because in practice, the data of a problem are often acquired progressively. It is obviously not reasonable to recompute the whole result each time new data is inserted. Thus (semi-)dynamic schemes are necessary. We study both theoretical complexity and practical efficiency of the algorithms. This thesis presents our work in this field.

In the first chapter, the definitions and main properties of classical structures in Computational Geometry are recalled, as well as a fundamental tool

—duality— whose presentation is new, and which allows us to prove well known results in a manner that is more intuitive than the usual one. All background necessary for understanding our algorithms will be found in this chapter.

Some previous results on randomized incremental algorithms are presented in Chapter 2. These algorithms are the ones that marked the beginning of research in this field. They are static, though incremental, because they all make use of the Conflict Graph, a structure that needs the knowledge of all data during the initialization step. It must be noticed that K.L. Clarkson also worked a lot on randomized divide-and-conquer algorithms, but these algorithms cannot be extended to a dynamic scheme, and thus will not be developed here. We also present several methods of analysis, due to K.L. Clarkson, K. Mulmuley and R. Seidel.

The Delaunay Tree is introduced in Chapter 3. It is, (pre-)historically(!), our first semi-dynamic structure [BT86, BT93]. It allows the construction of the Delaunay triangulation of a set of point sites, in any dimension, without preliminary knowledge of this set. The Delaunay Tree is compared to a similar structure proposed in 1990 by L.J. Guibas, D.E. Knuth, and M. Sharir.

Chapter 4 shows how the basic idea of the Delaunay Tree leads to designing a very general structure, the Influence Graph, allowing the semi-dynamic construction of numerous geometric structures. We can roughly say that this structure applies to the same problems for which the Conflict Graph was previously used, but it allows insertions.

Two kinds of complexity analysis are presented, which hold under hypotheses similar to those stated for analyzing algorithms using the Conflict Graph. The possibility of removing these conditions is also studied.

Several applications are developed. The use of the Influence Graph for efficiently responding to queries is also discussed.

Experimental results for computing convex hulls in dimensions 3 and 4, and Delaunay triangulations in dimensions 2 and 3, show the practical efficiency of the algorithms.

In Chapter 5, the idea of the Delaunay Tree is picked up again, and enlarged to be applied to the case of higher orders Voronoi diagrams, in any dimensions. This case does not fit into the framework of Chapter 4. Experimental results are shown.

Chapter 6 finally shows how to dynamize the Influence Graph. The general scheme for deleting data is simple, although the details are rather technical. We illustrate it on two examples: Delaunay triangulation of point sites and arrangements of line segments in the plane. The theoretical complexity is nevertheless excellent, and the algorithm is very efficient in practice.

At the same time as we were developing the Influence Graph, authors were designing other structures. A quick presentation of these structures will be found in Chapter 7.

Chapter 1

Fundamental structures

In this chapter, we present the definitions and main properties of fundamental structures in Computational Geometry, for which algorithms of constructions are presented in the following chapters. The complexity of the best known deterministic algorithms is also given. Most of the results are taken from [Ede87, Meh84, PS85], where more details can be found, especially concerning algorithms, that will not be studied here.

$I\!\!E^d$ denotes the d-dimensional Euclidean space, with scalar product $\langle \cdot, \cdot \rangle$, and Euclidean distance δ. Terms such as "vertical", "above", "below", will be often used : the vertical direction is parallel to the last axis of $I\!\!E^d$, and its orthogonal hyperplane is thus horizontal.

1.1 Convex hull

The *convex hull* of a finite set S of n points of $I\!\!E^d$, denoted as $E(S)$, can be equivalently defined in the following way as :

- the smallest convex set containing S,

- the intersection of all convex sets containing S,

- the intersection of all halfspaces containing S.

The computation of the convex hull is a central problem in Computational Geometry, and it has been vastly studied, not only because of its practical applications, but also because it is a preliminary for solving a lot of other questions.

Classical mathematical results for convex polyhedra allow us to state the following property (let us recall that a k-face is a face of dimension k) :

> *Property 1.1* The number of k-faces and the number of incidences between k-faces and $k+1$-faces of $E(S)$ are $O\left(n^{\min\left(\lfloor \frac{d}{2} \rfloor, k+1\right)}\right)$, for $0 \le k \le d - 1$. These bounds are asymptotically tight.

In particular, in both dimensions 2 and 3, this number is linear.

The behaviour of the complexity of $E(S)$ has also been studied under hypotheses for the distributions of the points of S. For example, if the points are chosen independently from a normal distribution, then the expected size of $E(S)$ is $O\left((\log n)^{\frac{d-1}{2}}\right)$.

The problem of computing the convex hull consists in computing the polyhedron that it forms, i.e. the complete description of its boundary.

Let us start with a first observation. In the plane, it is easy to see that this problem is at least as complex as sorting n numbers : let us take the points on a parabola with vertical axis, then the convex hull is the polygon formed by these n points, *in the order of abscissae*. Each algorithm computing the convex hull of n points has thus a $\Omega(n \log n)$ time complexity.

Optimal static algorithms exist in dimensions 2 and 3.

In higher dimensions, R. Seidel [Sei81] has given an algorithm whose complexity is $O\left(n\log n + n^{\lfloor\frac{d+1}{2}\rfloor}\right)$, which is optimal for even dimension ≥ 2. The storage is in $O\left(n^{\lfloor\frac{d}{2}\rfloor}\right)$. The algorithm is incremental, but the analysis is amortized on the n insertions.

After several years of unsuccessful researchs on finding an optimal algorithm for any dimensions, B. Chazelle has recently given such an algorithm [Cha91].

An algorithm whose complexity is $O(n^2 + f\log n)$, where f is the size of the output, that is the number of facets of the convex hull, is described in [Sei86].

In the plane, deterministic on-line algorithms are known, with optimal complexity $O(\log n)$ per insertion (see for example [AES85]).

1.2   Voronoi diagram

1.2.1   Voronoi diagram for points in $I\!\!E^d$

The *Voronoi diagram* of a set S of n points, called *sites*, in $I\!\!E^d$ is a geometric structure used to solved proximity queries, such as the closest neighbors of a given site, or the closest pair of sites.

This structure has been plentifully studied in the literature, and we started our works with it. F. Aurenhammer recently devoted a very complete survey to it [Aur91], in which he gives the applications, the properties, and a lot of algorithms for computing the diagram.

For each site p, the *Voronoi cell* $V(p)$ of p is the set of points in $I\!\!E^d$ that are closer to p than to any other site of S.

$$V(p) = \{x \in I\!\!E^d, \forall q \in S \setminus \{p\}, \delta(x,p) < \delta(x,q)\}$$

$V(p)$ can also be expressed as :

$$V(p) = \bigcap_{q \in S \setminus \{p\}} H(p,q)$$

where $H(p,q)$ is the halfspace limited by the bisecting hyperplane of p and q, containing p.

The Voronoi diagram of S is the partitioning of $I\!\!E^d$ formed by the Voronoi cells of the sites.

The points of S are assumed to be in general position : no $d+2$ points are cospherical. In this case, the vertices of the diagram are the centers of the spheres passing through $d+2$ sites.

> *Property 1.2* The complexity of the Voronoi diagram of a set of n points in $I\!\!E^d$ is $O\left(n^{\lceil\frac{d}{2}\rceil}\right)$.

In the 3 dimensional case, this complexity is quadratic.

The *Delaunay triangulation* is the dual graph of the Voronoi diagram : two sites of S are linked by an edge of the triangulation if and only if their respective cells are adjacent. The Delaunay triangulation is a partition of $I\!\!E^d$ in simplices whose vertices are the sites. It satisfies the following property, which defines it uniquely :

> The spheres circumscribing the simplices of the Delaunay triangulation do not contain any site in their interior.

This property allows us to design a very simple incremental algorithm. This algorithm is due to Green and Sibson [GS78] for the two dimensional case, and it has been generalized to any dimensions [Bow81]. As many incremental algorithms, it is not efficient in the worst case : its complexity is $O(n^2)$ in the planar case and $O\left(n^{\lceil \frac{d}{2} \rceil + 1}\right)$ in dimension d. The authors show that the complexity can be improved to an expected cost of $O\left(n^{1 + \frac{1}{d}}\right)$ for homogeneous distributions. However, the proof is not really formal, the hypotheses are not very precise and the performances tend to be poorer for degenerate distributions such as points on surfaces.

The algorithm consists in inserting the points in turn, and in updating the structure after each insertion. When a new point $\mathbf{m}$ is inserted, the simplices whose circumscribed sphere contains $\mathbf{m}$ must be removed, they do not belong to the triangulation any more. The union of these simplices is a simply connected region $R(\mathbf{m})$. If $F(\mathbf{m})$ denotes the set of facets of the boundary of $R(\mathbf{m})$, the new simplices are obtained by linking $\mathbf{m}$ to the elements of $F(\mathbf{m})$ (Figure 1.1). This algorithm is very simple, and allows us to progressively acquire new data, which is of great practical interest. To increase its efficiency, a data structure giving quickly the set $R(\mathbf{m})$ has to be found. This problem has been the starting point of our researches (Chapter 3).

Numerous optimal algorithms are known for the planar case. The recent algorithm by B. Chazelle, for the computation of convex hulls, now gives by duality (section 1.4) a worst-case optimal algorithm in any dimension.

When the points are uniformly distributed, the expected size of the Voronoi diagram is $O(n)$ [Dwy91]. In practice, in 3D space, for points on the surface of varied objects, its size is also often linear. A fundamental open question is to design *output-sensitive* algorithms, that is algorithms whose complexity is a function of the size t of the diagram. This has been solved in the particular case where the points lie in two planes [Boi88, BCDT91].

1.2.2 Higher order Voronoi diagrams

Planar case

$\mathcal{S}$ denotes a set of n sites in general position in the plane. Given $\mathcal{T}$ a subset of points of $\mathcal{S}$, the generalized Voronoi polygon of $\mathcal{T}$ is defined as :

$$V(\mathcal{T}) = \{\mathbf{p}, \forall \mathbf{v} \in \mathcal{T}, \forall \mathbf{w} \in \mathcal{S} \setminus \mathcal{T}, \delta(\mathbf{p}, \mathbf{v}) < \delta(\mathbf{p}, \mathbf{w})\}$$

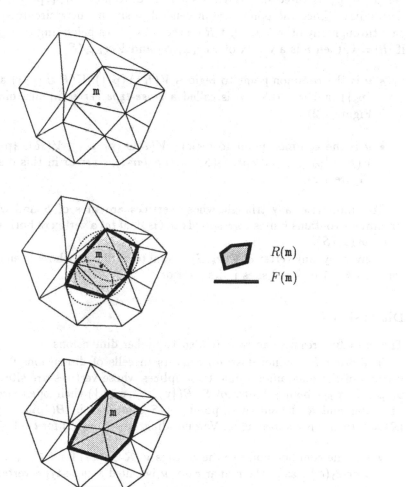

Figure 1.1 : Inserting a site in the Delaunay triangulation

$V(\mathcal{T})$ is the locus of a point p that is closer to each site of $\mathcal{T}$ than to any other point in $\mathcal{S} \setminus \mathcal{T}$. The *order k Voronoi diagram* of $\mathcal{S}$ is ($|\cdot|$ denotes cardinality)

$$Vor_k(\mathcal{S}) = \{V(\mathcal{T}),\ \mathcal{T} \subseteq \mathcal{S},\ |\mathcal{T}| = k\}$$

Let p_1, p_2, p_3 be three sites of $\mathcal{S}$, ν their circumcenter, $B(p_1, p_2, p_3)$ their open circumdisk. Since the points are in general position, their circumcircle do not pass through any other site. Let R be the set of sites belonging to $B(p_1, p_2, p_3)$. If $|R| = k$, then ν is a vertex of $Vor_{k+1}(\mathcal{S})$ and $Vor_{k+2}(\mathcal{S})$:

- ν is the common point to regions $V(R \cup \{p_1\})$, $V(R \cup \{p_2\})$ and $V(R \cup \{p_3\})$ in $Vor_{k+1}(\mathcal{S})$. ν is called a *close-type vertex* in this diagram (see Figure 1.2).

- ν is the common point to regions $V(R \cup \{p_1, p_2\})$, $V(R \cup \{p_2, p_3\})$ and $V(R \cup \{p_3, p_1\})$ in $Vor_{k+2}(\mathcal{S})$. ν is a *far-type vertex* in this diagram (see Figure 1.2).

To summarize, any triangle whose vertices are sites of $\mathcal{S}$ and whose open circumdisk contains k sites corresponds to (*is dual to*) a vertex of both $Vor_{k+1}(\mathcal{S})$ and $Vor_{k+2}(\mathcal{S})$.

Conversely, any vertex of $Vor_k(\mathcal{S})$ is dual to a triangle whose open circumdisk contains $k - 1$ or $k - 2$ sites in its interior.

Dimension d

The preceding results can be extended to higher dimensions.

The order k Voronoi diagram consists in cells of dimensions $0, \ldots, d$. the vertices of the diagram are dual to simplices whose vertices are sites of $\mathcal{S}$. Let $p_1, p_2, \ldots, p_{d+1}$ be $d + 1$ sites of $\mathcal{S}$, $B(\{p_1, \ldots, p_{d+1}\})$ their open circumball, ν its center and $\mathcal{R}$ the subset of points of $\mathcal{S}$ contained in $B(\{p_1, \ldots, p_{d+1}\})$. If $|\mathcal{R}| = l$, then ν is a vertex of all Voronoi diagrams $Vor_k(\mathcal{S})$ for $l + 1 \leq k \leq l + d$.

- ν is the common point to the regions $V(\mathcal{R} \cup \{p_i\})$ for $i \in \{1, \ldots, d+1\}$ in $Vor_{l+1}(\mathcal{S})$: as in the planar case, ν is called a close-type vertex.

- In $Vor_{l+h}(\mathcal{S})$, ν is the common point to the regions $V(\mathcal{R} \cup \mathcal{P})$ where $\mathcal{P}$ may be any subset of size h of $\{p_1, \ldots, p_{d+1}\}$; ν is incident to $\binom{d+1}{h}$ regions. If $1 < h < d$ we call ν a *medium-type vertex*.

- ν is the common point to the regions $V(\mathcal{R} \cup \{p_1, \ldots, p_{d+1}\} \setminus \{p_i\})$ for $i \in \{1, \ldots, d+1\}$ in $Vor_{l+d}(\mathcal{S})$: as in dimension 2, ν is a far-type vertex.

To summarize, a close-type vertex of a higher order Voronoi diagram remains in d successive diagrams. More generally, a h-face remains in $d - h$ successive diagrams.

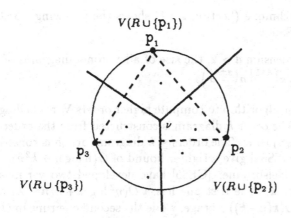

Close-type vertex in $Vor_{k+1}(S)$

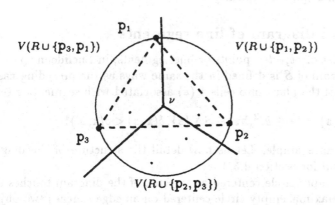

Far-type vertex in $Vor_{k+2}(S)$

Figure 1.2 : Two types of vertices

Complexity results

The order k Voronoi diagram has been introduced in [SH75] to deal with k closest points problems. Lee [Lee82] gives the following result :

> **Property 1.3** In the plane, the size of the order k Voronoi diagram is $O(k(n-k))$. The size of orders $\leq k$ Voronoi diagrams is thus $O(nk^2)$.

The random sampling technique (Section 2.3.1) shows the following worst-case result in dimension d [CS89] :

> **Property 1.4** In dimension $d > 2$, the size of all Voronoi diagrams of orders 1 to k is $O\left(k^{\lceil\frac{d+1}{2}\rceil}n^{\lfloor\frac{d+1}{2}\rfloor}\right)$.

Lee has also given the first algorithm to compute higher orders Voronoi diagrams for n sites in the plane. The order k diagram is constructed from the order $k-1$ diagram in $O(k(n-k)\log n)$ time. The orders $\leq k$ diagrams are thus constructed in $O(k^2 n \log n)$ time. [AGSS89] give a tighter bound of $O(n \log n + k^2 n)$.

B. Chazelle and H. Edelsbrunner [CE85] have developed two versions of a better algorithm for large k. The first one has a $O(n^2 \log n + k(n-k)\log^2 n)$ time complexity, with a $O(k(n-k))$ storage, while the second one runs in $O(n^2 + k(n-k)\log^2 n)$ time with $O(n^2)$ storage. H. Edelsbrunner, J. O'Rourke and R. Seidel [EOS86] use duality (Section 1.4) to construct all orders $\leq n-1$ diagrams in $O(n^{d+1})$ time and space complexity.

1.2.3 Voronoi diagram of line segments

S now denotes a set of *objects* : points or line segments, in Euclidean space E^2. The Voronoi diagram of S is defined in the same ways as the preceding case, it is the tesselation of the plane into cells $V(s)$ associated with segments $s \in S$:

$$V(s) = \{x \in E^2, \forall s' \in S \setminus \{s\}, \delta(x,s) < \delta(x,s')\}$$

Figure 1.3 shows an example. Let us now detail the structure of the diagram, which will be useful for Section 4.3.3.1.

The maximal empty circle centered at a vertex of the diagram touches three objects, and the maximal empty circle centered on an edge touches two objects.

An edge Γ of this diagram is part of the bisecting line of two objects p and q, and its two extremities are equidistant from p, q and r and from p, q and srespectively. Γ is defined by four segments and denoted as (pq, r, s). It consists of portions of lines and parabolas. To describe the case of infinite edges, we add the symbol ∞. The region (pq, r, ∞) corresponds to an unbounded part of the bisecting line of p and q (Figure 1.5). Moreover, to ensure connectivity, we add edges at infinity, that will bound unbounded cells : the edge $(p\infty, r, s)$ in Figure 1.5 is the set of centers (at infinity) of the circles (with infinite radius, that is lines) that touch p, and it is limited by the circles that respectively touch r and s. In Figures 1.4 and 1.5, the shadowed parts are the interiors of the empty circles centered on edge Γ.

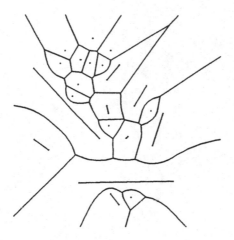

Figure 1.3 : Voronoi diagram of line segments

In some ambiguous cases, label (pq, r, s) may denote two different edges, that will be distinguished in the following way : $(pq, r, s)^+$ and $(pq, r, s)^-$ (Figure 1.6).

The geometric dual of the Voronoi diagram of line segments is the edge Delaunay triangulation. C.K. Yap [Yap87] gives an optimal algorithm running in $O(n \log n)$ to compute the Voronoi diagram of lines and portions of circles.

1.3 Arrangements

An *arrangement* is the partition of $I\!\!E^d$ induced by a finite set of hypersurfaces or portions of hypersurfaces.

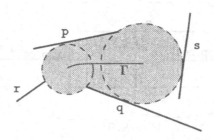

Figure 1.4 : Edge (pq, r, s)

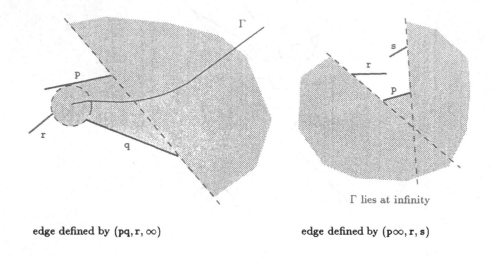

Γ lies at infinity

edge defined by (pq, r, ∞)   edge defined by (p∞, r, s)

Figure 1.5 : Infinite edges

1.3.1 Arrangements of hyperplanes

S is a set of hyperplans in $I\!\!E^d$. The general position condition sets that any $d+1$ hyperplanes have empty intersection, so a vertex of the arrangement belongs to exactly d hyperplanes. The arrangement consists of cells that are convex polyhedra.

An arrangement of n hyperplanes in $I\!\!E^d$ can be computed together with its incidence graph in optimal running time $O(n^d)$. The algorithm is incremental, it leans on the Zone Theorem [ESS91], that states that the insertion of an hyperplane can be achieved in $O\left(n^{d-1}\right)$.

The *k-level* of an arrangement is the set of points p in $I\!\!E^d$ such that the number of hyperplanes of S strictly above p is $\leq k - 1$ and the number of hyperplanes strictly below p is $\leq n - k$ (the hyperplanes are assumed to be non vertical).

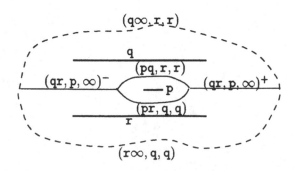

Figure 1.6 : Ambiguous label (pq, r, s)

Figure 1.7 shows the 2-level in an arrangement of lines in the plane.

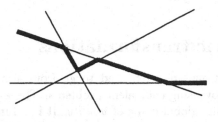

Figure 1.7 : 2-level in an arrangement of hyperplanes

K.L. Clarkson (see Section 2.3.1) has proved that :

Property 1.5 The size of levels 1 to k in an arrangement of n hyperplanes in $\mathbb{E}^d$ is $O\left(n^{\lfloor\frac{d}{2}\rfloor}k^{\lceil\frac{d}{2}\rceil}\right)$.

1.3.2 Trapezoidal map

In the case when S consists of n line segments in plane $\mathbb{E}^2$, the *trapezoidal map* of the arrangement is obtained by drawing vertical lines from the extremities and the intersection points of the segments, limited by the firstly encountered segment in both directions upwards and downwards (Figure 1.8).

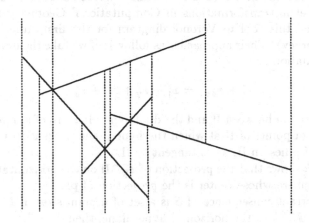

Figure 1.8 : Trapezoidal map

If a denotes the number of intersection points between the n segments, it can easily be seen that this map has $O(n+a)$ trapezoids, and that it completely describes the arrangement.

The best deterministic algorithm is due to B. Chazelle and H. Edelsbrunner [CE88], who compute the arrangement of n line segments in $O(n \log n + a)$ time and $O(n + a)$ storage.

1.4   Geometric transformations

Geometric transformations are often used to go from a structure to another, and to deduce results by using equivalent problems. These transformations are usually known under the global name of *duality*. It is a fundamental tool with different aspects that we describe more precisely in the following.

The origin of duality is mathematical *polarity*, that associates an hyperplane of $I\!\!E^d$ with a point of $I\!\!E^d$, and conversely : a point p of coordinates

$$(p_1, p_2, \ldots, p_d)$$

is dual to the hyperplane of equation

$$(\pi_{\mathbf{p}}) \quad x_d = 2p_1 x_1 + 2p_2 x_2 + \ldots + 2p_{d-1} x_{d-1} - p_d,$$

and vice versa. This transformation preserves incidence, and reverses the "above-below" relation. It easily implies that

> The convex hull of a set of points is dual to the intersection of a set of half-spaces.

Geometric transformations can use spaces of different dimensions : H. Edelsbrunner and F. Aurenhammer (and others) have worked a lot on results obtained owing to geometric transformations in Computational Geometry (see [Bro79] who has applied this tool to Voronoi diagram for the first time, [ES86], and [Aur91] for a survey). Their approach is as follows : if we take the unit paraboloid in $I\!\!E^{d+1}$, of equation

$$(\text{II}) \quad x_{d+1} = x_1^2 + x_2^2 + \ldots + x_d^2$$

then the intersection between II and the dual hyperplane $\pi_{\mathbf{p}}$ of a point p exterior to II is the set of points of II at which the hyperplane tangent to II contains p. In particular, if p lies on II, $\pi_{\mathbf{p}}$ is tangent to II.

This remark shows that the projection of $\pi_{\mathbf{p}} \cap$ II on the horizontal hyperplane $x_{d+1} = 0$ is a sphere whose center is the projection of p.

As an important consequence, if S is a set of n point sites in $I\!\!E^d$, and if $I\!\!E^d$ is immersed in $I\!\!E^{d+1}$ as the horizontal hyperplane, then

> The Voronoi diagram of S is the projection onto $I\!\!E^d$ of the 1-level (upper envelope) of the arrangement in $I\!\!E^{d+1}$ of the hyperplanes dual to the points obtained by projecting the sites of S onto II.

In fact, the intersection of two hyperplanes tangent to II at the two respectively projected points of p and q onto II, projects onto the horizontal hyperplane

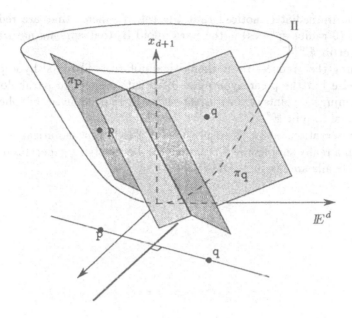

Figure 1.9 : Duality between Voronoi diagram and arrangement

$\mathbb{E}^d$ on the bisecting hyperplane of p and q (Figure 1.9). In the same way, the intersection point between $d+1$ hyperplanes of this arrangement, associated with $d + 1$ sites of $\mathcal{S}$, projects on the center ν of the sphere passing through these $d+1$ sites. The number of sites interior to the sphere is the number of associated hyperplanes in the arrangement, that pass above this intersecting point. More generally, we get :

> The order k Voronoi diagram of $\mathcal{S}$ is dual to the k-level of the arrangement in $\mathbb{E}^{d+1}$ of the hyperplanes dual to the sites of $\mathcal{S}$.

In [Bro79], K.Q. Brown proved the duality between Voronoi diagram and convex hull, using inversions. In [ES86, Aur91], the proofs of such properties where purely analytic. We propose in [DMT92b] a much more geometric interpretation, that avoids all calculations (see also [BCDT91]). We now give a quick clue of the reasonings obtained with this new point of view.

The core idea resides in the definition of the *space of spheres* $\mathcal{O}$ in $\mathbb{E}^d$. A sphere in $\mathbb{E}^d$ with equation

$$(\mathbf{p}) \quad \langle \mathbf{x}, \mathbf{x} \rangle - 2\langle \mathbf{x}, \Phi \rangle + \chi = 0$$

is represented by the point with coordinates

$$\mathbf{p} = (\Phi, \chi)$$

in $\mathcal{O}$. The space of spheres is isomorphic to $\mathbb{E}^{d+1}$.

It can be immediately noticed that the set of spheres that are reduced to their center (0 radius spheres) is the paraboloid Π, that appears naturally as a particular set in $I\!\!E^{d+1}$.

Everything then reduces to mathematical polarity. The dual hyperplane π_p is nothing else but the polar hyperplane of p with respect to paraboloid Π, or the set of conjugate points of p with respect to Π, that is the set of spheres that are orthogonal to p in $I\!\!E^d$.

These observations allow us to use very classical mathematical results and to deduce in a really simple way all duality results involving generalized Voronoi diagrams. Details are given in [DMT92b].

Chapter 2

Static randomized incremental algorithms

Several techniques to analyze randomized incremental algorithms are presented in this chapter, as well as a data structure, the Conflict Graph due to K.L. Clarkson. This structure sets the *static* character of the incremental algorithms using it : if a new data is introduced, there is no way of updating the result, the new result must be computed by starting the algorithm again from the beginning.

A formalization of the problem, borrowed from [CS89], except in a few details, is firstly described. This general formalism can be applied to numerous geometric problems. The notations introduced here will be used throughout the following chapters.

This chapter does not aim at exhaustively citing the results that can be obtained with randomized static incremental algorithms, but to present the essentials of the most important ideas. These ideas (particularly those of K.L. Clarkson, who pioneered the field) have given rise to an impressive quantity of papers.

2.1 Formalization of the problem

All the geometric problems are formulated in very general terms.

The data of the problem are *objects*, elements of an universe $\mathcal{O}$. The considered objects are subsets of the working space, for instance the points, the line segment, the lines... in the Euclidean space $\mathbb{E}^d$ of dimension d.

The *regions* belong to an universe $\mathcal{F}$, also consisting of subsets of the working space.

We take interest in the regions defined by the objects. If $\mathcal{S}$ is a set of objects, we denote as $\mathcal{S}^{(b)}$ the set of subsets of $\mathcal{S}$ having at most b elements, for an integer $b \in \mathbb{N}^*$; let ∇ be a relation between $\mathcal{F}$ and $\mathcal{S}^{(b)}$. We say that $F \in \mathcal{F}$ is *defined* by the set $\mathcal{S}$ if there is an element $X \in \mathcal{S}^{(b)}$ such that $F \nabla X$. We say in this case that X *determines* F. $\mathcal{F}(\mathcal{S})$ denotes the set of regions defined by $\mathcal{S}$.

$$\mathcal{F}(\mathcal{S}) = \{F \in \mathcal{F} \ / \ (\exists X \in \mathcal{S}^{(b)}) \ F \nabla X\}$$

The relation ∇ is assumed to be *functional*, i.e. for each region $F \in \mathcal{F}(\mathcal{S})$, the subset X of $\mathcal{S}$ determining F is unique. This formalizes the usual hypotheses of *general position* on the set of objects in $\mathcal{S}$.

With each region is associated its *influence range* that is a subset of the universe of objects $\mathcal{O}$. The objects belonging to the influence range of a region F are said to be *in conflict* with F. By definition, the influence range of F does not contain any of the objects determining F. The definition of an influence range, thus of a conflict, will of course be given precisely for each application.

For $F \in \mathcal{F}$ and $\mathcal{S} \subset \mathcal{O}$, $\mathcal{S}$ finite, we denote as $\mathcal{S}(F)$ the set of objects of $\mathcal{S}$ in conflict with F, and we call *width* of F with respect to $\mathcal{S}$ the number $|\mathcal{S}(F)|$.

It will be assumed at the moment (a more general problem will be studied in Chapter 5) that the problem can be expressed as :

> *Compute the regions defined by $\mathcal{S}$ and without any conflict with the objects in $\mathcal{S}$ (regions with zero width).*

Let us illustrate these definitions on the **example of the Delaunay triangulation**. In $I\!\!E^d$, the universe $\mathcal{O}$ is the set of points of $I\!\!E^d$. The universe of regions is the set of simplices and halfspaces of $I\!\!E^d$. If S denotes a set of point sites, a region can be determined by d or $d+1$ sites. Let $b = d+1$, $\mathcal{F}(S)$ is the set of simplices with its $d+1$ vertices belonging to S and halfspaces whose boundary is a hyperplane containing d points of S.

The influence range of a simplex is its open circumball, and the influence range of a halfspace is its own interior. ∇ is functional if and only if any $d+2$ sites are never cospherical and $d+1$ sites never lie on a common hyperplane. To compute the Delaunay triangulation of S is exactly to compute the set of regions without conflict defined by S.

In the **example of the intersection of halfspaces**, the objects are halfspaces of $I\!\!E^d$, and the regions are edges of the intersection, determined by $d+1$ objects. An edge and a halfspace are in conflict if the edge is not contained in the closed halfspace. The intersection of the halfspaces is the set of edges without conflict.

The **example of the intersections of a set of line segments** will also be studied in the sequel. The algorithm constructs a trapezoidal map (Section 1.3.2). An object (a segment line) is in conflict with a region (a trapezoid) if the segment intersects the trapezoid. A trapezoid is defined by at most 4 segments. The set of empty trapezoids gives the arrangement of the set of line segments.

Additional notations will be necessary.

If j is an integer, we denote as $\mathcal{F}_j(S)$ the set of regions of $\mathcal{F}(S)$ of width j with respect to S, and as $\mathcal{F}_{\leq j}(S)$ the set of regions of width at most j. $\mathcal{F}_j^{(i)}(S)$ is the subset of $\mathcal{F}_j(S)$ consisting in the regions determined by elements X in $S^{(b)}$ with cardinality $|X| = i \leq b$. $\mathcal{F}_{\leq j}^{(i)}(S)$ is defined in a similar way.

All preceding notations can be defined in the same manner for any subset $\mathcal{R}$ of S. The width of a region defined by $\mathcal{R}$ might be calculated with respect to any set of objects. For the sake of brevity, we call "region of width j (with no more precision) defined by $\mathcal{R}$" a region defined by $\mathcal{R}$ and of width j with respect to $\mathcal{R}$. If the algorithm is incremental, and if $\mathcal{R}$ is the set of present objects at a given time ($\mathcal{R}$ is the current set of objects), the *current width* of a region defined by S is its width with respect to $\mathcal{R}$.

We already define the notations connected to random sampling. A subset $\mathcal{R}$ of S is a *random sample* of S if the elements of $\mathcal{R}$ are randomly chosen from the elements of S. All random samples of size r of S are equally likely, they have probability $\frac{1}{\binom{n}{r}}$ where n denotes the cardinality of S.

We denote as $f_j(r, S)$ the mathematical expectation $E_{(r,S)}[|\mathcal{F}_j(\mathcal{R})|]$ of the number $|\mathcal{F}_j(\mathcal{R})|$ of regions of width j defined by a random sample $\mathcal{R}$ of size r of S. In the same way, $f_j^{(i)}(r, S) = E_{(r,S)}[|\mathcal{F}_j^{(i)}(\mathcal{R})|]$. Finally, $\phi_j(r, S)$ is the maximum of the expectations $f_j(r', S)$ for $1 \leq r' \leq r$.

The notations defined that have been defined in this section are summarized in Figure2.1.

$\mathcal{O}$ .... universe of objects

$\mathcal{F}$ .... universe of regions

$\mathcal{S}$ .... set of objects $\mathcal{S} \subset \mathcal{O}$

$\mathcal{S}^{(b)}$ .... set of all subsets of $\mathcal{S}$ of cardinality at most b

$\mathcal{F}(\mathcal{S})$ .... set of all regions defined by $\mathcal{S}$

$\mathcal{S}(F)$ .... set of all objects of $\mathcal{S}$ in conflict with region F

$\mathcal{F}_j(\mathcal{S})$ .... set of all regions of $\mathcal{F}(\mathcal{S})$ of width j

$\mathcal{F}_{\leq j}(\mathcal{S})$ .... set of all regions of $\mathcal{F}(\mathcal{S})$ of width $\leq j$

$\mathcal{F}_j^{(i)}(\mathcal{S})$ .... set of all regions of $\mathcal{F}_j(\mathcal{S})$ defined by i objects

$\mathcal{F}_{\leq j}^{(i)}(\mathcal{S})$ .... set of all regions of $\mathcal{F}_{\leq j}(\mathcal{S})$ defined by i objects

$f_j(r, \mathcal{S})$ .... expectation of $|\mathcal{F}_j(\mathcal{R})|$ for $\mathcal{R} \subset \mathcal{S}$ of cardinality r

$f_j^{(i)}(r, \mathcal{S})$ .... expectation of $|\mathcal{F}_j^{(i)}(\mathcal{R})|$ for $\mathcal{R} \subset \mathcal{S}$ of cardinality r

$\phi_j(r, \mathcal{S})$ .... maximum of $f_j(r', \mathcal{S})$ for $r' \leq r$

Figure 2.1 : Notations

2.2   A data structure : the conflict graph

The general idea leading to the conflict graph of K.L. Clarkson is not specific to geometric problems.

The principle of the conflict graph is illustrated in [CS89] on the example of insertion sort. The insertion sort of n numbers can be seen as an alternative to divide-and-conquer for quicksort. At each step, a new element is put into its place. If we maintain a sorted list of the already inserted elements, this sort can be time-consuming, because a large proportion of the sorted list may be examined at each step.

Two main ideas are used to speed up this algorithm :

- Let us assume that we now, for each uninserted value, its location in the current list, and conversely for each location, the list of corresponding uninserted values. These informations constitute the conflict graph. When element c is inserted, if its location is between a and b, it can be directly inserted. We then have to update in the conflict graph the set of uninserted values between a and b, by comparing them with c. The insertion time is thus proportional to the number of elements between a and b.

- If in addition, the numbers are inserted in a randomized fashion, that is in random order, at step r, the already inserted values are fairly evenly distributed among the whole set to be sorted. The number of values between a and b is about $\frac{n}{r}$ on the average.

These two hypotheses allow us to conclude that the expected randomized complexity is

$$O\left(\sum_{r=1}^{n}\frac{n}{r}\right) = O(n\log n)$$

This summarizes the scheme of incremental randomized algorithms using the conflict graph. It is to be noticed that in those algorithms, the insertion cost is dominated by the time of updating the informations about conflicts.

The formalization of this scheme can be done in the framework of Section 2.1. The geometric problem reduces to constructing the set $\mathcal{F}_0(\mathcal{S})$ of regions with width zero defined by a set $\mathcal{S}$ of n objects.

The incremental algorithm consists in adding the objects of $\mathcal{S}$ one by one to the current subset $\mathcal{R}$ of $\mathcal{S}$ while maintaining the set $\mathcal{F}_0(\mathcal{R})$ of regions without conflict defined by $\mathcal{R}$.

At step r, when object o is added to the current subset $\mathcal{R}$, the set of regions having current width zero is updated by removing the regions of $\mathcal{F}_0(\mathcal{R})$ in conflict with o and adding new regions having current width zero : the regions defined by subsets of $\mathcal{R} \cup \{o\}$ containing o, and without conflict with any object of $\mathcal{R} \cup \{o\}$.

To speed up this updating, the algorithm maintains, in addition to the set of regions $\mathcal{F}_0(\mathcal{R})$, the *conflict graph*. It is a bipartite graph, defined on the cartesian product $\mathcal{F}_0(\mathcal{R}) \times \mathcal{S} \setminus \mathcal{R}$, with an edge for each pair (F, o) of a region F of $\mathcal{F}_0(\mathcal{R})$ and an object o in $\mathcal{S} \setminus \mathcal{R}$ in conflict with F.

The conflict graph thus quickly gives (in time linear in the number of desired regions) the regions of $\mathcal{F}_0(\mathcal{R})$ conflicting with the new object o studied in the current step. Per contra, each incremental stage must now include a phase updating the conflict graph. The edge of the conflict graph incident to the regions of $\mathcal{F}_0(\mathcal{R})$ conflicting with o are removed, and new edges are created to represent the conflicts involving the new regions of current width zero (regions in $\mathcal{F}_0(\mathcal{R} \cup \{o\}) \setminus \mathcal{F}_0(\mathcal{R})$) and the remaining objects (objects in $\mathcal{S} \setminus (\mathcal{R} \cup \{o\})$). The complexity of each step is thus at least proportional to the number of edges in the conflict graph that are updated (created or removed) at that step.

The definition of the conflict graph imposes the knowledge of the whole set $\mathcal{S}$ from the initialization step. That is why the algorithms using it are intrinsically static, though incremental. Our goal in the following chapters will be to introduce a structure quite as general, but allowing us to break free from this constraint.

2.3   Techniques of analysis

2.3.1   Random sampling

A first idea

P. Erdös and J. Spencer [ES74] have been the first ones to write a whole book devoted to probabilistic techniques. Their aim was to make the scientific community know these powerful techniques, that had been applied to a wide variety of combinatorial problems with much success.

This book contains new proofs of already known theorems, in various fields on combinatorics, mainly graph theory. The technique consists in replacing "counting arguments" by "random variable arguments", in nonconstructive proofs of existence.

Let us consider one of the first examples in the book.

A *tournament* T on a set V of vertices is a directed graph (set of couples of V^2) in which, for each pair $\{x, y\} \subset V$, such that $x \neq y$, either $(x, y) \in T$, or $(y, x) \in T$ holds, but not both. Let T_n be the set of tournaments on $V_n = \{1, \ldots, n\}$ for a given integer n. An *hamiltonian path* P in a tournament $T \in T_n$ is an element of the set Σ_n of permutations of V_n, such that for all $i \in \{1, \ldots, n-1\}$, $(P(i), P(i+1)) \in T$ holds. The following result is due to T. Szele[1].

Theorem [Szele] *There is some $T \in T_n$ containing at least $n! \, 2^{-n+1}$ hamiltonian paths.*

Proof. by T. Szele

Let $U = \{(T, P)/T \in T_n, P \in \Sigma_n \text{ hamiltonian path in } T\}$

The cardinality of T_n is $2^{\binom{n}{2}}$.

If $P \in \Sigma_n$ is fixed, the number of tournaments T such that $(T, P) \in U$ is $2^{\binom{n}{2} - (n-1)}$ since P determines $n - 1$ edges of T. Since there are $n!$ permutations of V_n, U has $|U| = n! \, 2^{\binom{n}{2} - (n-1)}$ elements.

On the other hand, $|U| = \sum_{T \in T_n} |\{P \in \Sigma_n / (T, P) \in U\}|$.

This sum has $2^{\binom{n}{2}}$ terms, so there is some $T \in T_n$ such that

$$|\{P \in \Sigma_n / (T, P) \in U\}| \geq n! \, 2^{-(n-1)}$$

$\square$

Proof. by P. Erdös and J. Spencer

Let $\mathbf{T}$ be a random variable with values in T_n, such that

$$\forall T \in T_n, \operatorname{Prob}(\mathbf{T} = T) = 2^{-\binom{n}{2}}$$

or, equivalently,

$$\forall i, j \in V_n, \operatorname{Prob}((i, j) \in \mathbf{T}) = \frac{1}{2}$$

with independent probabilities for each pair $\{i, j\}$.

For a fixed $P \in \Sigma_n$ let us define a function f_P on T_n by

$$f_P(T) = \begin{cases} 1 \text{ if } P \text{ is an hamiltonian path in } T \\ 0 \text{ otherwise} \end{cases}$$

[1]Kombinatorikai vizsgálatok az irányított teljes graffál kapcsolatban, *Mat. Fiz. Lapok* 50, 1943, 223-256

$h(T) = \sum\limits_{P \in \Sigma_n} f_P(T)$ is the number of hamiltonian paths in T.

The additivity of mathematical expectation gives :

$$
\begin{aligned}
E[h(T)] &= E\left[\sum_{P \in \Sigma_n} f_P(T)\right] \\
&= \sum_{P \in \Sigma_n} E[f_P(T)] \\
&= \sum_{P \in \Sigma_n} 2^{-n+1} \\
&= n!\, 2^{-n+1}
\end{aligned}
$$

Thus, there necessarily is some $T \in \mathcal{T}_n$ such that

$$h(T) \geq n!\, 2^{-n+1}$$

□

Application to computational geometry

K.L. Clarkson has used this idea to build a technique for proving combinatorial results in computational geometry. In his numerous papers, he applies it firstly to specific algorithms —post-office problem, convex hulls, intersections of line segments— [Cla85, Cla87, CS88], then he defines, together with P.W. Shor, a very general formalism [CS89] whose essentials are repeated in Section 2.1.

We use the notations defined in Section 2.1.

K.L. Clarkson and P.W. Shor bound the number $|\mathcal{F}^{(i)}_{\leq k}(S)|$ of regions with width at most k defined by a set S of n objects. We will give here the proof of this theorem because it is representative of all proofs using random sampling. It uses two fundamental lemmas.

Lemma 2.1 *Let S be a set of n objects and F a region determined by i objects of S and having width j with respect to S ($F \in \mathcal{F}^{(i)}_j(S)$). Then the probability $p^{(i)}_{j,k}(r)$ that F be a region of width k defined by a random subset of size r of S is :*

$$p^{(i)}_{j,k}(r) = \dfrac{\dbinom{j}{k} \dbinom{n-i-j}{r-i-k}}{\dbinom{n}{r}}$$

Proof. Let $\mathcal{R}$ be a random subset of size r of S. There are $\binom{n}{r}$ possibilities to choose $\mathcal{R}$. Region F in $\mathcal{F}^{(i)}_j(S)$ belongs to $\mathcal{F}^{(i)}_k(\mathcal{R})$ if the i objects determining it belong to $\mathcal{R}$, and if in addition F is in conflict with k objects among the j objects of S in conflict with it. $\mathcal{R}$ then must contain $r - i - k$ objects chosen from the $n - i - j$ remaining objects of S. □

Lemma 2.2 *Let S be a set of n objects and $\mathcal{R}$ a random sample of size r of S. Then the mathematical expectation $f_k^{(i)}(r, S)$ of the number of regions determined by i objects of $\mathcal{R}$ and having width k with respect to $\mathcal{R}$ is given by :*

$$f_k^{(i)}(r, S) = \sum_{j=0}^{n-i} |\mathcal{F}_j^{(i)}(S)| \frac{\binom{j}{k}\binom{n-i-j}{r-i-k}}{\binom{n}{r}}$$

Proof. $f_k^{(i)}(r, S)$ is a function of the probability for a region $F \in \mathcal{F}(S)$, determined by i objects, to belong to $\mathcal{F}_k^{(i)}(\mathcal{R})$:

$$
\begin{aligned}
f_k^{(i)}(r, S) &= \sum_{j=0}^{n-i} \sum_{F \in \mathcal{F}_j^{(i)}(S)} \mathrm{Prob}(F \in \mathcal{F}_k^{(i)}(\mathcal{R})) \\
&= \sum_{j=0}^{n-i} |\mathcal{F}_j^{(i)}(S)|\, p_{j,k}^{(i)}(r)
\end{aligned}
$$

Lemma 2.1 allows us to conclude.   □

We can prove the following theorem, owing to those two lemmas :

Theorem 2.3 [Clarkson-Shor] *Let S be a set of n objects. Then the number $|\mathcal{F}_{\leq k}(S)|$ (for $k \geq 2$) of regions having width at most k defined by S is bounded by a function of the mathematical expectation $f_0\left(\lfloor \frac{n}{k} \rfloor, S\right)$ of the number of regions without conflict defined by a random sample of size $\frac{n}{k}$ of S, as follows :*

$$|\mathcal{F}_{\leq k}(S)| \leq 4(b+1)^b k^b f_0\left(\lfloor \tfrac{n}{k} \rfloor, S\right)$$

(b still denotes, as in the definition of $\mathcal{F}(S)$, the maximal number of objects determining a region)

Proof. The inequality $|\mathcal{F}_{\leq k}^{(i)}(S)| \leq 4(b+1)^i k^i f_0^{(i)}\left(\lfloor \tfrac{n}{k} \rfloor, S\right)$ is proved for each value of i between 1 and b.

From Lemma 2.2, the mathematical expectation $f_0^{(i)}(r, S)$ of the number of regions without conflict defined by a random sample of size r of S is :

$$
\begin{aligned}
f_0^{(i)}(r, S) &= \sum_{j=0}^{n-i} |\mathcal{F}_j^{(i)}(S)| \frac{\binom{n-i-j}{r-i}}{\binom{n}{r}} \\
&\geq \sum_{j=0}^{k} |\mathcal{F}_j^{(i)}(S)| \frac{\binom{n-i-j}{r-i}}{\binom{n}{r}}
\end{aligned}
$$

Calculations on factorials show that, for $j \leq k$,

$$\frac{\binom{n-i-j}{r-i}}{\binom{n}{r}} \geq \frac{r(r-1)\ldots(r-i-1)}{n(n-1)\ldots(n-i-1)} \left(\frac{n-r-k+1}{n-i-k+1}\right)^k.$$

For $r = \lfloor \frac{n}{k} \rfloor$ and $k \geq 1$, we have in addition

$$\frac{n-r-k+1}{n-i-k+1} \geq 1 - \frac{1}{k},$$

thus we can give the lower bound :

$$\left(\frac{n-\lfloor \frac{n}{k} \rfloor - k + 1}{n-i-k+1}\right)^k \geq \frac{1}{4} \text{ for } k \geq 2.$$

Comparing with the inequality obtained for $f_0^{(i)}(r, S)$, we get :

$$|\mathcal{F}_{\leq k}^{(i)}(S)| \leq 4 \frac{n(n-1)\ldots(n-i-1)}{\lfloor \frac{n}{k} \rfloor \left(\lfloor \frac{n}{k} \rfloor - 1\right)\ldots\left(\lfloor \frac{n}{k} \rfloor - i - 1\right)} f_0^{(i)}\left(\lfloor \frac{n}{k} \rfloor, S\right)$$

and using $\frac{n}{k} - 1 < \lfloor \frac{n}{k} \rfloor$, we obtain finally :

$$|\mathcal{F}_{\leq k}^{(i)}(S)| \leq 4k^i \frac{1}{\left(1 - \frac{bk}{n}\right)^i} f_0^{(i)}\left(\lfloor \frac{n}{k} \rfloor, S\right)$$

from which we deduce the result by bounding k above by $\frac{n}{b+1}$. □

Remark 2.4 This theorem only applies for k between 2 and $\frac{n}{b+1}$. The following bounds can nevertheless be deduced from it :

$$|\mathcal{F}_0(S)| \leq |\mathcal{F}_{\leq 1}(S)| \leq |\mathcal{F}_{\leq 2}(S)|$$
$$\leq 4(b+1)^b 2^b f_0\left(\lfloor \frac{n}{2} \rfloor, S\right)\left(1 + O\left(\frac{1}{n}\right)\right)$$

On the other hand, for k close to n, we have the trivial bound :

$$|\mathcal{F}_{\leq k}(S)| \leq |\mathcal{F}(S)| = O(n^b)$$

The theorem gives interesting bounds when an upper bound on $f_0\left(\lfloor \frac{n}{k} \rfloor, S\right)$ is known. Let us take the example of a set S of hyperplanes in $\mathbb{E}^d$. The theorem shows, as announced in Section 1.3.1 :

Corollary 2.5 [Clarkson-Shor] *For any set of hyperplanes in general position in $\mathbb{E}^d$, and for any integer $k \geq 2$, the size of all orders $\leq k$ levels in the arrangement is bounded by $O\left(n^{\lfloor \frac{d}{2} \rfloor} k^{\lceil \frac{d}{2} \rceil}\right)$.*

and another result already mentioned in Section 1.2.2 :

Corollary 2.6 [Clarkson-Shor]
    For any set of points in general position in $\mathbb{E}^d$, and for any $k \geq 2$, the size of the orders $\leq k$ Voronoi diagrams is bounded by $O\left(n^{\lfloor\frac{d+1}{2}\rfloor}k^{\lceil\frac{d+1}{2}\rceil}\right)$.

    Let us now consider an incremental algorithm. This algorithm is randomized if the objects are introduced at random, that is, at each step, all objects have the same probability to be chosen. In this case, the set $\mathcal{R}$ of present objects at a given stage is a random sample of $\mathcal{S}$.

    As seen in Section 2.2, if a conflict graph is used, the complexity of each step is at least proportional to the number of edges of the conflict graph created or removed at that step. We impose the following condition :

    Update condition : at each incremental step, updating the set of regions having current width zero, and the conflict graph, can be done within a time proportional to the number of edges of the conflict graph created or removed at that step.

    The fundamental theorem on complexity of randomized algorithms is presented below. Its proof is given in [CS89], it uses random sampling in a similar way as the proof of Theorem 2.3. The expected time for inserting an object at step $r+1$ is shown to be

$$O(\phi_0(r,\mathcal{S}))\frac{n-r}{r^2}.$$

($\phi_0(r,\mathcal{S})$ is the maximum of $f_0(r',\mathcal{S})$ for $r' \leq r$)

Theorem 2.7 [Clarkson-Shor] *A randomized incremental algorithm using a conflict graph and satisfying the update condition runs on a set of objects $\mathcal{S}$ of size n in expected time*

$$O\left(n\sum_{r=1}^{n}\frac{\phi_0(r,\mathcal{S})}{r^2}\right).$$

So, for example

- if $\phi_0(r,\mathcal{S}) = O(r)$, then the expected running time of the algorithm is $O(n\log n)$.

- if $\phi_0(r,\mathcal{S}) = O(r^\alpha)$ with $\alpha > 1$, then the expected complexity is $O(n^\alpha)$.

Let us mention two applications for this problem :

Theorem 2.8 [Clarkson-Shor] *Let S be a non degenerate set of n line segments in the plane, among which a pairs intersect. The a intersections can be computed with a randomized incremental algorithm with expected $O(a + n\log n)$ running time and $O(n+a)$ storage.*

Theorem 2.9 [Clarkson-Shor] *A randomized incremental algorithm computes the intersection of n halfspaces of $\mathbb{E}^d$ in expected $O(n\log n)$ running time for $d \leq 3$ and $O\left(n^{\lfloor\frac{d}{2}\rfloor}\right)$ for $d \geq 4$.*

2.3.2   Probabilistic games and Θ series

In parallel with K.L. Clarkson, K. Mulmuley worked on randomized incremental algorithms. His analyses are totally different and do not use random sampling. His results are nevertheless similar, and the data structure is also a conflict graph. He firstly studied arrangements of curves in the plane [Mul88, Mul89a] then arrangements of hyperplanes in space and higher orders Voronoi diagrams [Mul89b, Mul91a].

In [Mul91a], the k first levels in an arrangement of a set S of n hyperplanes in $I\!E^d$ are computed. $v(l)$ denotes the size of the l-level. Let us define

$$\Theta_{l_0}(s, j, S) = \sum_{l=1}^{l_0} v(l) + \sum_{l=l_0+1}^{j} \left(\frac{l_0}{l}\right)^s v(l)$$

The complexity of the algorithm is given as a function of $\Theta_k(d, n, S)$.

The probabilistic model used for drawing the hyperplane out in random order is a sequence of independent Bernoulli trials : for each hyperplane of S, it independently tosses a coin having probability $\frac{1}{r}$ of success. The selected hyperplanes are those for which success occurred. $\Phi_r^{l_0}$ denotes the number of vertices of the levels of orders at most l_0 in the arrangement of the so obtained random sample. We then compute the expectation $E[\Phi_r^{l_0}]$ in two following ways.

On one hand, a simple reasoning is done (similar to the one used in the proof of Lemma 2.1), permitting to say that a vertex v with level $l - 1$ appears, for $l \leq l_0$, if the d hyperplanes determining it are drawn out, and in addition, for $l > l_0$, if at most $l_0 - 1$ hyperplanes among the $l - 1$ below v are drawn out. This leads to an expression of $E[\Phi_r^{l_0}]$. On the other hand, owing to the chosen probabilistic model that gives usual mathematical tools such as Γ function or Chernoff bound, $E[\Phi_r^{l_0}]$ can be bounded (some generalized zone theorem, proved in the same paper, is also used).

Comparing the two results gives bounds on Θ series, from which the following result follows :

Theorem 2.10 [Mulmuley] *The k first levels in an arrangement of hyperplanes in $I\!E^d$ can be computed in $O\left(k^{\lceil \frac{d}{2} \rceil} n^{\lfloor \frac{d}{2} \rfloor}\right)$ expected running time for $d \geq 4$ ($O\left(kn \log \frac{n}{k}\right)$ for $d = 2$ and $O\left(k^2 n \log \frac{n}{k}\right)$ for $d = 3$) using a randomized incremental algorithm.*

The Voronoi diagrams of orders 1 to k are obtained by the same method in $O\left(k^{\lceil \frac{d+1}{2} \rceil} n^{\lfloor \frac{d+1}{2} \rfloor}\right)$ expected running time for $d \geq 3$ ($O(nk^2 + n \log n)$ for $d = 2$).

K. Mulmuley's recent work on dynamic randomized algorithms are presented in Section 7.3.2.

2.3.3   Backwards analysis

This method gives very nice and elegant proofs, because they are surprisingly simple. [Sei91c] is one of the first papers in which this idea was published.

R. Seidel imputes to P. Chew the authorship of this technique that remained in the background for a long time. P. Chew might have used it for the first time to analyze an algorithm computing the Delaunay triangulation of a convex polygon, an algorithm that had never been published before R. Seidel [Sei91a] or O. Devillers (who also uses a similar analysis that is presented in Chapter 4) [Dev92] wrote it.

The method consists in analyzing the algorithm as if it was running backwards in time, from output to input. It does not use any formalism, nor any specific calculation technique.

Let us study the classical example of computing the convex hull of a set S of n point sites. A conflict graph is maintained between the edges of the current convex hull and the uninserted points. Let us describe the algorithm recursively computing the convex hull of $R \subset S$:

If $|R| = 3$, then we obtain the triangle spanned by the 3 points of R.

Else, choose a random point in R, recursively compute the convex hull of $R' = R \setminus \{q\}$ and insert q in the following way :

If q is contained in the convex hull of R', there is nothing more to do. Otherwise, the conflict graph gives a visible edge from q, and a traversal of the convex hull gives all edges visible from q, since they form a chain. Then it is sufficient to replace that chain with the two edges that have q as an endpoint.

The cost of updating the convex hull is proportional to the number of edges removed by inserting q, that can be large. However, since a removed edge will never appear again, this cost can be charged on the creation of these edges. But, at each insertion, at most 2 edges are created, so the total cost of updating the hull is linear.

The cost of updating the conflict graph must be evaluated. It is enough to maintain, for each uninserted point p, the edge E_p of the convex hull crossed by segment $[cp]$, where c is a fixed point interior to the hull (for example the circumcenter of the 3 first points). Conversely, for each edge E, we maintain the set of points p for which $E = E_p$.

The cost of updating the conflict graph when q is inserted is proportional to the number of points p for which E_p changes. For given p, E_p changes if, after inserting q (here comes the backwards analysis), q is an endpoint of E_p. As q is randomly chosen from R, the probability that q be chosen as one of the endpoints of E_p is $\frac{2}{r}$, where $r = |R|$. The expected number of changes for p is thus at most $\frac{2}{r}$ (if p already lies in R, there is no change). If we sum for all values of $r \leq n$, we get that the total number of changes for p is at most $O(\log n)$.

Since initializing the conflict graph takes a linear time, the whole cost of the algorithm is $O(n \log n)$.

This example shows the simplicity of the method : no calculations or complicated reasoning was needed.

Conclusion

We presented previous works on incremental randomized algorithms. A common point to all these algorithms is that the results hold whatever the point distribution might be, contrary to more classical analyses such as in [Dwy91]. The only condition is that the points must be inserted in random order.

These algorithms are all static because they use the conflict graph. This structure has been successfully used by a lot of authors, to compute various geometric structures. Let us mention [MMO91] who compute abstract Voronoi diagrams, [Mul88, Mul89a] who constructs arrangements in the plane, [Mul89b, Mul91a] who studies levels in arrangements of hyperplanes, and of course [CS89] who develop numerous applications. One of the aims of following chapters is to design a fully dynamic structure.

Let us notice that, in the theorems proved up to now, the cost of inserting an object is dominated by the cost of updating the conflict graph, that is not directly a function of the size of the change in the desired structure. We can roughly say that the cost of inserting the first objects is high, while the change in the output is of constant size.

Numerous other results have been obtained in parallel or after ours. Some of these results will be summarized in the chapters presenting our work, and the other ones together form Chapter 7.

Chapter 3

The Delaunay Tree

The Delaunay Tree is a hierarchical data structure which is defined from the Delaunay triangulation and, roughly speaking, represents a triangulation as a hierarchy of balls. It allows an "on-line", or semi-dynamic construction of the Delaunay triangulation of a finite set of n points in any dimensions : the points are not supposed to be known in advance, as in Chapter 2. On the contrary, the Delaunay Tree allows the data to be given while the incremental algorithm is running, because it makes possible to look for the conflicts of the point to be inserted, without storing them explicitly.

To this aim, the idea of maintaining the *history* of the construction —the successive versions of the triangulation, linked together in the tree— is introduced for the first time in [BT86]. The history allows a hierarchical location in the current triangulation, which is very efficient. The update of the structure is easy. This idea has been then often used again in randomized dynamic algorithms (see Sections 3.3 and Chapter 7). However, K. Mulmuley created a totally different dynamic data structure (see Section 7.3.2).

We only present the algorithm and the data structure in this chapter. The analysis will be given in Section 4.3.3.1, as a consequence of the general analysis of the Influence graph (see also [BT93]).

A similar structure, introduced afterwards by L.J. Guibas, D.E. Knuth and M. Sharir [GKS92], is presented at the end of this chapter.

3.1 Structure

The nodes of the tree are associated with the successive simplices of the triangulation. The word "simplex" will denote the geometric object as well as the corresponding node.

For the initialization step we choose $d + 1$ sites of S. They generate one finite simplex and $d + 1$ *infinite* ones (see Figure 3.1) : a halfspace will also be called a simplex, and will be considered as having d finite vertices and one vertex at infinity. The ball circumscribing an infinite simplex is exactly the halfspace itself, that can be considered as a ball with infinite radius. These $d + 2$ simplices will be the sons of the root of the tree.

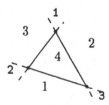

Figure 3.1 : Initialization step

Then the other points are inserted one after another. The triangulation is updated as indicated in Section 1.2.1. As already defined in Section 2.1, we say that a site p is in conflict with a simplex if p lies in the interior of it circumscribing ball.

After the insertion of site p, the simplices in conflict with p disappear from the triangulation, but remain in the Delaunay Tree. They are called *dead* and p is their *killer*. Some of those simplices have a facet on the boundary $F(\mathbf{p})$ of the region $R(\mathbf{p})$ formed by the simplices in conflict with p. Let T be one of the simplices in conflict with p that has a facet F belonging to $F(\mathbf{p})$. We construct a new simplex S, *created* by p, having vertex p and facet F. Let N be the simplex sharing facet F with T before the insertion of p.   Because the triangulation is

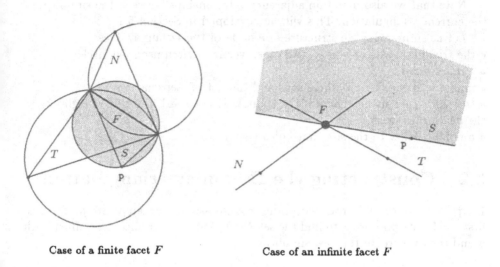

Case of a finite facet F Case of an infinite facet F

Figure 3.2 : Insertion of p

a Delaunay one, we have the following property (see Figure 3.2) :

> *Property 3.1* The circumscribing ball of S is included in the union of the two balls circumscribing T and N

This property is fundamental for the correctness of the algorithm, as will be seen in the sequel.

The newly created simplex S will be called : *son of T* and *stepson of N* through facet F. These edges will never been modified during the construction.

When it is killed, a simplex receives from 0 (if it has no facet on $F(\mathbf{p})$) to $d+1$ sons. As long as it is not dead, it can on the other hand receive an arbitrary number of stepsons : when a new point $\mathbf{p}'$ will be inserted, in conflict with S but not with N, S will be killed in turn, and its son S' having vertex $\mathbf{p}'$ and facet F will be another stepson of N. Thus a node has at most one son and one list of stepsons through each facet, that is : 0 to $d+1$ sons and 0 to $d+1$ *lists* of stepsons. The following property can nevertheless be used to bound the size of the structure :

> *Property 3.2* The total size of the stepson lists in the Delaunay Tree
> is less than the number of nodes, since each newly created node (the
> $d + 2$ sons of the root excepted) has exactly one stepfather. This is
> true in any dimension.

This hierarchical structure is called a *Delaunay Tree* for short, but it is more exactly a rooted direct acyclic graph. This graph contains a tree : the tree whose links are the links between fathers and sons.

A simplex of the current triangulation is not dead, and so corresponds to a node having no son, but possibly stepsons.

Note that we also maintain adjacency relationships between the simplices of the current triangulation. This will be developed in Section 5.1.3.5.

Let us summarize the structure of a node of the Delaunay Tree :
- the triangle : creator vertex, two other vertices, circumscribed circle
- a mark **dead**
- pointers to the at most three sons and the list of stepsons
- the three current neighbors if the triangle is not dead, the three neighbors at the death otherwise
- a pointer **killer** to the site that killed the triangle

3.2 Constructing the Delaunay triangulation

Let **p** be a site to be introduced in the triangulation. Two steps are performed: first, we locate **p** in order to find the set $R(\mathbf{p})$ of all the simplices in conflict with **p** and then we create the new simplices.

3.2.1 Location

If **p** is in conflict with a simplex T, Property 3.1 implies that **p** is in conflict with the father or the stepfather of T (or both of them). So we will be able to find all the simplices which are killed by **p** by recursively exploring the Delaunay Tree. Procedure location given in Figure 3.3 describes this traversal.

Procedure location(p,T) :

> **if** T has not been visited yet
> and **p** is in conflict with T **then**
>> **for** each stepson S of T location(p,S) ;
>> **for** each son S of T location(p,S) ;
>> **if** T is not dead **then**
>>> mark T killed by **p** ;
>>> add T to the list $R(\mathbf{p})$ of the killed simplices.

Figure 3.3 : Location of a site in the Delaunay Tree

Remark 3.3 It would also be possible to stop the recursive traversal as soon as a first simplex of the current triangulation in conflict with p is found. The other simplices in conflict with p would then be obtained by following neighborhood relations (see the end of Section 3.1).

Remark 3.4 All edges between a simplex and its successive stepsons may be traversed by Procedure location. Figure 3.4 shows an example. Points 1 to 7 are numbered in insertion order. The subgraph

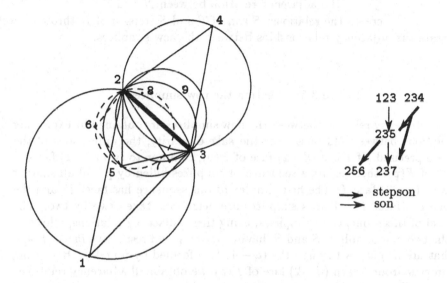

Figure 3.4 : All stepsons may be useful

of the Delaunay Tree corresponding to triangles 123, 234, 235, 256 et 237 is shown on the right side of the figure.

If site 8 is now introduced, it is in conflict with 235 and 256. but not with 123. The edges traversed to locate 8 are those from 234 to 235 and from 235 to 256. No path passing through 237 leads from 234 to 235.

If we want to locate 9, which is in conflict with 237, but not with 235, we must use the link from 234 to 237.

Thus, it is necessary to store both stepsons of 234.

3.2.2 Creating the new simplices

We go through the list $R(p)$ of the killed simplices. A facet F of a simplex T is on $F(p)$ if the simplex N neighbor of T through F is not killed. In this case, we create the simplex S with vertex p and facet F, and the two edges between S, its father T and its stepfather N. Moreover N and S are neighbors through F. Figure 3.5 describes Procedure creation realizing these updatings.

Procedure creation(p):

    for each simplex T killed by **p**
        for each neighbor N of T through a facet F
            if p is not in conflict with N **then**
                create the simplex S having vertex **p** and facet F ;
                replace the adjacency relation between N and T
                    by the adjacency relation between N and S ;
                create the relations: S son of T and S stepson of N through facet
    create the adjacency relationships between the new simplices.

Figure 3.5 : Creating the new simplices

The adjacency relations between the new simplices are obtained by exploiting the relations between old ones. For the sake of clarity, the dimensions of the faces are precised. If f is a $(d-2)$-face of $F(\mathbf{p})$, common to two $(d-1)$-faces F and F' of $F(\mathbf{p})$, there exists a sequence of simplices killed by **p** and all sharing the same $(d-2)$-face f. The first simplex of this sequence has facet F, and the last one has facet F'. Theses simplices are neighbors, taken two by two. The traversal of this sequence of simplices, using their adjacency relations, allows to link the two new simplices S and S' having vertex **p** and respective facets F and F', that are neighbors through the $(d-1)$-face formed by **p** and f. Repeating this process around each $(d-2)$-face of $F(\mathbf{p})$, we obtain all adjacency relations. Figure 3.6 illustrates this.

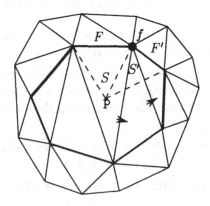

Figure 3.6 : Adjacency relations between new simplices

Each killed simplex has at most $d+1$ $(d-1)$-faces on $F(\mathbf{p})$, each of them has d $(d-2)$-faces, and each $(d-2)$-face is shared by two $(d-1)$-faces. A killed simplex is thus traversed at most $\frac{d(d+1)}{2}$ times during this search for neighbors among the new simplices.

The analysis of the algorithm is presented in section 4.3.3.1. Let us only give the result stated in Proposition 4.21 :

> *The Delaunay triangulation of n points in d-space can be computed on-line with $O\left(n^{\lfloor\frac{d+1}{2}\rfloor}\right)$ expected space in dimension $d \geq 2$.*
>
> *The expected update time for an insertion is $O(\log n)$ if $d = 2$ and $O\left(n^{\lfloor\frac{d+1}{2}\rfloor - 1}\right)$ if $d \geq 3$. These results are optimal.*

3.3   Another structure

L.J. Guibas, D.E. Knuth and M. Sharir proposed another structure [GKS92], also based on the history of the construction.

The construction is incremental, but the updating of the triangulation is different. It relies on the technique of switching the diagonal inside the quadrilateral composed by two adjacent triangles : if two adjacent triangles pqr and rsp are given, if s is in conflict with pqr, then pqr and rsp are replaced by pqs and qrs.

When a new site p is inserted, it must be located, which means here that we look for the triangle abc of the Delaunay triangulation containing p, and not those whose circumscribing circle contains p. Then the triangulation is updated as follows : first, abc is replaced by pab, pbc, pca ; then there is a propagation in the triangulation by examining the triangles by pairs of adjacent triangles, and possibly exchanging the diagonals.

The structure contains all triangles, Delaunay triangles and non Delaunay triangles, created at any time of the construction. Edges are created between a killed triangle and the new triangles partially covering it. In this way, the lists of stepsons are avoided, and the number of edges outcoming from a triangle is constant. but some triangles that have never been Delaunay triangles remain in the structure. The randomized analysis of this algorithm, based on random sampling, yields, despite these differences, to the same complexity as the Delaunay Tree (however, the analysis that is presented in [GKS92] is amortized).

A major drawback of this technique was, until a very recent time, that it was based on an updating process (the exchange of diagonals), that had never been generalized to dimensions greater than 2. But V.T. Rajan [Raj91] has just filled this lack, which might allow the extension of this structure to higher dimensions.

Conclusion

We have defined a semi-dynamic data structure for constructing the Delaunay triangulation, based on the knowledge of the history of the incremental construction.

It is possible to generalize this structure, keeping the same idea of remembering the previous steps of the construction. This general structure, the Influence graph, is presented in Chapter 4.

Chapter 4

A general structure : the Influence Graph

The Influence Graph [BDSTY92] generalizes the Delaunay tree. It is presented here under the formalism and notations introduced in Section 2.1 (see Figure 2.1 on Page 25). A randomized analysis of the algorithms using this structure is proposed in Section 4.1.1. We also describe a different analysis, due to O. Devillers. Then, in Section 4.2, we study the efficiency of the Influence Graph for locating a site.

Section 4.3 develops several applications to semi-dynamic constructions (allowing insertions) : arrangements, Delaunay triangulations of point sites in $\mathbb{E}^d$, convex hull in $\mathbb{E}^d$, Voronoi diagrams of segments in the plane. In the case of convex hulls and Delaunay triangulations, some experimental results are given.

The last section of the chapter is devoted to some remarks about our results.

4.1 The general framework

The different algorithms presented here are incremental and introduce objects one by one. Let S be the set of objects which have already been introduced. At a given stage, the incremental algorithm inserts a new object in S and updates the set $\mathcal{F}_0(S)$ of regions of current zero width defined by S (recall that a region is determined by at most b objects). This is performed through the maintenance of a dynamic structure called the influence graph (I-DAG for short) described below.

The I-DAG is a rooted directed acyclic graph whose nodes are associated with regions that, at some stage of the algorithm, have appeared as regions of zero width defined by the set of objects that have been introduced at that stage. Although the I-DAG is not strictly speaking a tree, we speak of leaves, fathers, sons etc. in the obvious way. The nodes of the I-DAG associated with regions of $\mathcal{F}_0(S)$ are marked. When a new object o is added to S, one new node is created for each region in $\mathcal{F}_0(S \cup \{o\})$ that is not a region of $\mathcal{F}_0(S)$.

As in the Delaunay Tree, the already existing nodes are never deleted from the I-DAG but possibly unmarked (they are *dead*). The new nodes are linked by edges to nodes already present in the I-DAG. These edges are constructed in such a way that a new object o can be efficiently located in the structure ; locating o means here to traverse all the nodes whose associated regions are in conflict with o. When a new node corresponding to a region F of $\mathcal{F}_0(S \cup \{o\})$ is added to the I-DAG, we put edges between already existing nodes and the new nodes so that the influence range of F is included in the union of the ranges of its parents.

The I-DAG structure is characterized by the two following fundamental properties :

Property 1 At each stage of the incremental process, the regions of current zero width ($\mathcal{F}_0(S)$) are leaves of the I-DAG.

Property 2 The influence range of the region associated with a node is included in the union of the ranges of its parents.

The construction of the I-DAG can be sketched as follows:

- We initialize S with the b first objects. A node of the I-DAG is created for each region of $\mathcal{F}_0(S)$ and made son of the root of the I-DAG. The influence range associated with the root is the whole objects universe $\mathcal{O}$.

- At each subsequent step, a new object o is added to S and the I-DAG is updated. The two following substeps are performed:

  location substep. This substep finds all the nodes of the I-DAG whose regions have zero width and are in conflict with o. This is done by traversing every path from the root of the I-DAG down to the first node which is not in conflict with object o.

  creation substep. From the information collected during the first substep, the creation substep creates a new node for each region in $\mathcal{F}_0(S \cup \{o\}) \setminus \mathcal{F}_0(S)$ and links the new nodes to already existing nodes in the structure so that Properties 1 and 2 still hold. The details of this substep depend on each particular application.

4.1.1   Randomized analysis of the I-DAG

This subsection aims at providing a randomized analysis of the space and time required to build the I-DAG structure. Randomization concerns here only the order in which the inserted objects are introduced in the structure. Thus, if the current set of objects is a set S of cardinality n, our results are expected values that correspond to averaging over the $n!$ possible permutations of the inserted objects, each one being equally likely to occur.

The main results of this chapter are stated in Theorems 4.3 and 4.13 below, they give quite general upper bounds on the size and the update time of the I-DAG as functions of the expected size of the output for a sample of the input.

For the sake of clarity, we will make some hypotheses that simplify somewhat the analysis. These conditions are fulfilled by a large class of geometric problems and allow one to express the results in a simple way. However these hypotheses are not really necessary and will be removed in Section 4.1.4.

The update conditions

We first assume that three update conditions are satisfied. These conditions are almost equivalent to Clarkson's. As already mentioned these hypotheses are mainly for clarity and will be removed later.

(1) The number of sons of a node of the I-DAG is bounded.

(2) Given a region F and an object o, the test to decide whether or not o is in conflict with F can be performed in constant time.

(3) If the new object o added to the current set S is found to be in conflict with k regions of $\mathcal{F}_0(S)$ then the creation substep requires $O(k)$ time.

Expected storage

Lemma 4.1 *If S has cardinality n, the expected size of the I-DAG of S is :*

$$O\left(\sum_{j=1}^{n} \frac{f_0(\lfloor \frac{n}{j} \rfloor, S)}{j}\right)$$

Proof. The expected number of nodes $\eta(S)$, in the I-DAG of S can be obtained by summing, for all the regions F of $\mathcal{F}(S)$, the probability that F occurs as a node in the I-DAG, which is $\frac{i!j!}{(i+j)!}$ (the i objects determining F must be inserted before the j objects in conflict with F).

So,

$$\begin{aligned}
\eta(S) &= \sum_{i=1}^{b} \sum_{j=0}^{n-i} |\mathcal{F}_j^{(i)}(S)| \frac{i!j!}{(i+j)!} \\
&= \sum_{i=1}^{b} |\mathcal{F}_0^{(i)}(S)| \\
&\qquad + \sum_{i=1}^{b} \sum_{j=1}^{n-i} \left(|\mathcal{F}_{\leq j}^{(i)}(S)| - |\mathcal{F}_{\leq (j-1)}^{(i)}(S)| \right) \frac{i!j!}{(i+j)!} \\
&= \sum_{i=1}^{b} \sum_{j=1}^{n-i-1} |\mathcal{F}_{\leq j}^{(i)}(S)| \, i \, \frac{i!j!}{(i+j+1)!} \\
&\qquad + \sum_{i=1}^{b} |\mathcal{F}_{\leq (n-i)}^{(i)}(S)| \frac{i!(n-i)!}{n!} \\
&= O\left(\sum_{j=1}^{n} \frac{f_0(\lfloor \frac{n}{j} \rfloor, S)}{j}\right) \quad \text{by Theorem 2.3.}
\end{aligned}$$

According to Update condition 1, this bound applies also to the size of the I-DAG. □

Expected time

Lemma 4.2 *Under the update conditions, if S has cardinality n, the expected time for inserting the last object in the I-DAG is*

$$O\left(\frac{1}{n} \sum_{j=1}^{n} f_0(\lfloor \frac{n}{j} \rfloor, S)\right)$$

Proof. Under Update condition 2, the computing time spent to locate the last inserted object o is proportional to the total number of nodes of the I-DAG visited when locating o. Due to Update condition 1, the number of nodes visited when locating o is at most proportional to the number of nodes of the I-DAG associated with regions in conflict with o. Thus the expected time for locating the last inserted object o is proportional to the expected number, $\theta(\mathcal{S})$, of nodes of the I-DAG associated with regions in conflict with o.

Let F be a region of $\mathcal{F}_j^{(i)}(\mathcal{S})$. F is a region in conflict with o associated with a node of the I-DAG, if o is one of the j objects in conflict with F (thus the choice, for o, of any of the j objects among the n is possible) and if the i objects defining F have been inserted before the j objects in conflict with F. This occurs with probability $\frac{i}{n}\frac{i!(j-1)!}{(i+j-1)!}$. The expected number $\theta(\mathcal{S})$ of nodes visited during the last insertion is then obtained by summing, for all the regions F of $\mathcal{F}(\mathcal{S})$, the above probability. Using Theorem 2.3, this yields :

$$\theta(\mathcal{S}) = \sum_{i=1}^{b}\sum_{j=1}^{n-i}|\mathcal{F}_j^{(i)}(\mathcal{S})|\frac{i!j!}{n(i+j-1)!} = O\left(\frac{1}{n}\sum_{j=1}^{n}f_0(\lfloor\tfrac{n}{j}\rfloor,\mathcal{S})\right),$$

from a calculation similar to the proof of Lemma 4.1

Due to Update condition 3, the computing time of the last creation substep is also dominated by a term proportional to the number of nodes of the I-DAG associated with a region in conflict with o and admits the same expected upper bound as $\theta(\mathcal{S})$.   □

Main theorem

Lemmas 4.1 and 4.2 prove the main result of this section :

Theorem 4.3 *If the set of already inserted objects S has cardinality n and if the update conditions are fulfilled, the I-DAG of S requires*

$$O\left(\sum_{j=1}^{n}\frac{f_0(\lfloor\tfrac{n}{j}\rfloor,\mathcal{S})}{j}\right)$$

expected memory space. The insertion of the nth object can be done in

$$O\left(\frac{1}{n}\sum_{j=1}^{n}f_0(\lfloor\tfrac{n}{j}\rfloor,\mathcal{S})\right)$$

expected update time.

Corollary 4.4 *Under the update conditions, the total expected time to build a I-DAG for a set S of n objects is :*

$$O\left(\sum_{j=1}^{n}\phi_0(\lfloor\tfrac{n}{j}\rfloor,\mathcal{S})\right)$$

Proof. Notice that in that corollary, S is no longer the current subset of inserted objects, but the final set of objects. From Theorem 4.3, we know that, if a subset $\mathcal{R}$ of S with cardinality r has been inserted at a given time, the expected time to insert the last object of $\mathcal{R}$ is : $O\left(\sum_{j=1}^{r} \frac{1}{r} f_0(\lfloor \frac{r}{j} \rfloor, \mathcal{R})\right)$. This expected time accounts for averaging over the $r!$ permutations of the objects of $\mathcal{R}$. Now, the expected time to insert the rth object of S results from further averaging over the r-random samples of S which yields : $O\left(\sum_{j=1}^{r} \frac{1}{r} f_0(\lfloor \frac{r}{j} \rfloor, S)\right)$. The proof of Corollary 4.4 results from summing over r from 1 to n, knowing that $\phi_0(r, S)$ is a non decreasing function (recall that $\phi_0(r, S)$ is the maximum of the values $f_0(r', S)$, for $r' \leq r$). $\square$

Two immediate consequences of Theorem 4.3 are the following :

- If $f_0(x, S) = \Theta(x)$, the space complexity is $O(n)$ and the time to insert a new object is $O(\log n)$.

- If $f_0(x, S) = \Theta(x^{\alpha})$ (with $\alpha \geq 1$), the space complexity is $O(n^{\alpha})$ and the time to insert a new object is $O(n^{\alpha-1})$.

4.1.2   Influence graph versus Conflict graph

This result is the same as the complexity obtained by using the conflict graph, which is

$$O\left(n \sum_{r=1}^{n} \frac{\phi_0(r, S)}{r^2}\right)$$

(see Theorem 2.7).

In fact, $\lfloor \frac{n}{j} \rfloor = r$ for j such that $\frac{n}{r+1} < j \leq \frac{n}{r}$. By changing $\frac{n}{j}$ to r in our result, and grouping partial sums (we sum together all terms corresponding to values of j yielding the same $\lfloor \frac{n}{j} \rfloor$), yields the expression of K.L. Clarkson and P.W. Shor.

Another, maybe more pleasant, way to see the similarity is to look precisely at the total number of nodes traversed in the I-DAG in order to locate an object, during the whole construction : it is the sum, over all regions ever created, of the number of objects in conflict with this region. In other words, it is exactly the number of Conflict graph edges ever created.

The only difference between the two results (independently from the on-line possibility in the I-DAG) lies in the fact that the non amortized complexity of an insertion is totally different : the work done in the Conflict graph is more expensive for the first objects than for the last ones, while it is the contrary in the I-DAG.

4.1.3   Another analysis

The above analysis of the algorithms constructing the I-DAG uses random sampling to bound the number of regions in conflict with at most k objects, and

deduces time and space bounds for the algorithms. O. Devillers proposes in
[Dev92] a simple analysis, using the principle of backwards analysis presented in
Section 2.3.3. He avoids the use of random sampling techniques. He also analyzes the Conflict Graph in a similar manner, and applies his results to the design
of efficient algorithms combining Conflict and Influence graphs (see Section 7.1).

Lemma 4.5 *The expected value of the number of conflicts of the lth object with
regions that had no conflict at stage k (k < l) is $\frac{f_1(k+1,S)}{k+1}$.*

> *Proof.* Let ω be one of the $n!$ possible orderings on S. By averaging
> over ω, the first k objects plus the lth object o may be any sample
> of size $k + 1$ with the same probability. Suppose thus that o is
> introduced immediately after the first k elements. Then the regions
> that were defined by those k elements and that are in conflict with o,
> are determined by a subset of those $k + 1$ elements and their width
> is 1 after the insertion of o. Finally, o may be any element of the
> sample of size $k + 1$, with probability $\frac{1}{k+1}$, which yields the result. □

Lemma 4.6 *The expected number of regions created by the insertion of the kth
object is less than $\frac{bf_0(k,S)}{k}$.*

> *Proof.* A region created by the kth object o is a region defined by
> these k objects and without any conflict with them. In the same way
> as done in the preceding lemma, we can say that the first k objects
> may be any sample of S of size k with the same probability. Then $\frac{b}{k}$
> is an upper bound for the probability that o be one of the at most b
> objects determining a region.   □

Lemma 4.7 *The expected value of the number of conflicts of the lth object
with the regions created by the insertion of the kth object (k < l) is less than
$\frac{b}{k}\frac{f_1(k+1,S)}{k+1}$.*

> *Proof.* Let us denote $E_{k,l}$ this expected value. $E_{k,l}$ is the sum, over
> all regions F defined by S, of the probability $p_{F,k,l}$ that F be created
> by the insertion of the kth object and in conflict with the lth object
> o.
>
> Let us decompose this probability $p_{F,k,l}$, conditionally with the event
> that F is in conflict with o and had no conflict at stage k. In the case
> where F had no conflict at stage k, the probability that the object
> that created F be the last among the k objects is less than $\frac{b}{k}$, as in
> the preceding lemma. In the case where F had a conflict at stage k,
> F could not have been created by the kth object, so this case does
> not contribute to the value of $p_{F,k,l}$. So we conclude by using Lemma
> 4.5, which gives the expectation of the conditional event.   □

Theorem 4.8 *The complexity of the operations on the Influence graph are the
following :*

(1) *The expected size of the Influence graph at stage k is less than $\sum_{j=0}^{k} \frac{b f_0(j, \mathcal{S})}{j}$*

(2) *The expected cost of inserting the lth object in the Influence graph is less than $\sum_{j=0}^{l-1} \frac{b}{j} \frac{f_1(j+1, \mathcal{S})}{j+1}$*

(3) *The expected cost of inserting the lth object in the Influence graph, knowing the conflicts of this object with the regions that had no conflict at stage k, is less than $\sum_{j=k}^{l-1} \frac{b}{j} \frac{f_1(j+1, \mathcal{S})}{j+1}$*

The result *(3)* is the central point that allows to build accelerated algorithms (Section 7.1).

Proof.

(1) Since the number of sons is bounded, the size of the influence graph is proportional to its number of nodes. This number is simply the sum over all the regions of the probability for a region to be a node of the graph. By Lemma 4.6 the expected number of nodes created at stage j is less than $\frac{b f_0(j, \mathcal{S})}{j}$.

(2) During the insertion of the lth object, the conflicts are located by a traversal of the influence graph. A node F is visited, and yields a positive answer, if it is in conflict with the lth object. By summing over the stage of creation j of F we get $\left(\sum_{j=1}^{l-1} E_{j,l} \right)$. According to update conditions, the number of visited nodes, with positive answer, in the influence graph is linearly related to the cost of the insertion.

(3) Same result starting the summation at $j = k$.

$\square$

In the applications $f_0(r, \mathcal{S})$ and $f_1(r, \mathcal{S})$ are often both linear. In such a case, the complexities get a more explicit expression stated in the following theorem. Furthermore, if a direct expression of $f_1(r, \mathcal{S})$ is not available, it is possible to show (see [CS89], Theorem 2.3 and Remark 2.4) that $f_1(r, \mathcal{S}) = O\left(f_0\left(\lfloor \frac{r}{2} \rfloor, \mathcal{S}\right)\right)$, so it is sufficient to suppose that $f_0(r, \mathcal{S})$ is linear.

Theorem 4.9 *If $f_0(r, \mathcal{S}) = O(r)$,*

(1) *The expected size of the influence graph at stage k is $O(k)$.*

(2) *The expected cost of inserting the lth object in the influence graph is $O(\log l)$.*

The whole cost of the algorithm is $O\left(\sum_{l=1}^{n} \log l\right) = O(n \log n)$.

(3) *The expected cost of inserting the lth object in the influence graph, knowing the conflicts of this object with the regions that had no conflict at stage k, is $O\left(\log \frac{l}{k}\right)$.*

Proof. This theorem is simply a corollary of Theorem 4.8. $\square$

4.1.4   Removing the update conditions

In some cases the three update conditions can be removed. We will show in Section 4.3 that removing the update conditions (especially Condition 1, that is not verified in the case of the Delaunay Tree) will lead, in some cases, to simpler algorithms.

Constant test time and linear update time

Update conditions 2 and 3 can be relaxed. If the time required to check conflict between a region and an object exceeds $O(1)$, the overcost will simply appear as a multiplicative factor in the overall complexity. If the time required by the creation substep surpasses a linear function of the number of conflicts, it is in general not difficult to charge the overcost to the overall complexity. This kind of analysis will be done for example for the incremental construction of higher order Voronoi diagrams (see Section 5.1.3.5).

Bounded number of sons

It is more interesting to attempt to remove Update condition 1. The preceding analysis works because all the relevant quantities can be expressed as functions of the number of nodes. If the condition is not fulfilled, we must count the number of edges of the I-DAG and to that purpose, we introduce the notion of bicycles. A *bicycle* is a pair of regions of $\mathcal{F}(\mathcal{S})$ occurring as a father and one of its sons in the I-DAG associated with at least one permutation of the object set $\mathcal{S}$.

Notice that the maximum number of objects defining a bicycle is at most $2b$ and thus is still bounded. An object is in conflict with a bicycle if it is not any of the objects that determine the bicycle and if it is in conflict with at least one of the two regions forming the bicycle. Thus the influence range of a bicycle is the union of the influence ranges of the two regions forming the bicycle, excepting the objects defining these regions.

In analogy with the notation used for regions, the additional notations $\mathcal{G}(\mathcal{S})$ (set of bicycles defined by $\mathcal{S}$), $\mathcal{G}_j(\mathcal{S})$ (set of bicycles defined by $\mathcal{S}$ of width j), $\mathcal{G}_{\leq j}(\mathcal{S})$ (for bicycles of width at most j)... are naturally derived. We define $g_0(r,\mathcal{S})$ to be the expected size of $\mathcal{G}_0(\mathcal{R})$ for r-random samples $\mathcal{R}$ of $\mathcal{S}$.

With these definitions, the following lemma can be proved, using the random sampling technique, in a way similar to Theorem 2.3.

Lemma 4.10

$$|\mathcal{G}_{\leq j}(\mathcal{S})| \;=\; O\left(j^{2b} g_0(\lfloor \tfrac{n}{j} \rfloor, \mathcal{S})\right)$$

We can now compute the expected storage required by the I-DAG :

Lemma 4.11 *If $\mathcal{S}$ has cardinality n, the expected size of the I-DAG of $\mathcal{S}$ is :*

$$O\left(\sum_{j=1}^{n} \frac{g_0(\lfloor \tfrac{n}{j} \rfloor, \mathcal{S})}{j}\right)$$

Proof. The size of the I-DAG is linearly related to the number of its edges. A necessary condition for a given bicycle G to occur as an edge in the I-DAG is that the i objects defining G are inserted before the j objects in conflict with G (i.e. in conflict with one of the two regions associated with G). So the probability that $G \in \mathcal{G}(\mathcal{S})$ arises in the I-DAG is less than $\frac{i!j!}{(i+j)!}$. Calculations analogous to those appearing in the proof of Lemma 4.1 yield the result. $\square$

Lemma 4.12 *The expected time for inserting the nth object in the I-DAG is*

$$O\left(\frac{1}{n}\sum_{j=1}^{n}g_0(\lfloor\tfrac{n}{j}\rfloor,\mathcal{S})\right)$$

Proof. Let o be the nth object to be inserted. The number of times a node is visited is equal to the number of its parents which are in conflict with o. Thus the number of performed tests is no more than the number of bicycles in conflict with o occurring in the I-DAG.

Let G be a bicycle of $\mathcal{G}_j^{(i)}(\mathcal{S})$. G is in conflict with o and occurs as an edge in the I-DAG if the following two necessary conditions are satisfied : the i objects defining G are inserted before the j objects in conflict with G, and one of the j objects is o. The probability that G is in conflict with o is thus no more than $\frac{i}{n}\frac{i!(j-1)!}{(i+j-1)!}$. The result is then achieved as in Lemma 4.2. $\square$

Lemmas 4.11 and 4.12 prove the main results of this section, generalizing Theorem 4.3 and Corollary 4.4 :

Theorem 4.13 *If the set of already inserted objects $\mathcal{S}$ has cardinality n and if Update conditions 2 and 3 are fulfilled (but not Update condition 1), the I-DAG of $\mathcal{S}$ requires*

$$O\left(\sum_{j=1}^{n}\frac{g_0(\lfloor\tfrac{n}{j}\rfloor,\mathcal{S})}{j}\right)$$

expected memory space. The insertion of the nth object can be done in

$$O\left(\frac{1}{n}\sum_{j=1}^{n}g_0(\lfloor\tfrac{n}{j}\rfloor,\mathcal{S})\right)$$

expected update time.

Corollary 4.14 *Under Update conditions 2 and 3, the total expected time to build a I-DAG for a set $\mathcal{S}$ of n object is :*

$$O\left(\sum_{j=1}^{n}\gamma_0(\lfloor\tfrac{n}{j}\rfloor,\mathcal{S})\right)$$

where $\gamma_0(r,\mathcal{S})$ denotes the maximum value of $g_0(r',\mathcal{S})$ for $r' \leq r$.

4.2   Locating with the influence graph

4.2.1   Faster object location

If the following additional property is verified, it is possible to get a better complexity result for the search of a single region in conflict with a new object.

Property 3 The influence range of the region associated to a node is included in the union of the ranges of its children.

If Property 3 holds, then a conflict with a given new object can be found by following a simple path from the root of the I-DAG to a leaf.

Theorem 4.15 *If Property 3 holds then a conflict with any new object can be found in $O(\log n)$ time provided that the n objects that were inserted in the I-DAG have been inserted in random order.*

> *Proof.*   Let o be the new object and F be a region on the path from the root to a leaf of the I-DAG in conflict with o. Suppose that F has zero width after the insertion of the kth object. If F is determined by i objects, the conditional probability that F has been created during the insertion of the kth object is $\frac{i}{k} \leq \frac{b}{k}$. Indeed, for F to be created at step k, the kth inserted object must be one of the i objects defining F. (It is important to notice that the above probability is conditioned by the fact that F has width zero after the insertion of the first k objects.)
>
> Averaging this probability over the $\binom{n}{k}$ possible subsets of k objects introduced first in the I-DAG yields a probability less than $\frac{b}{k}$ that the node on the path would change after the insertion of the kth object. Thus the number of visited nodes is less than $\sum \frac{b}{k} = O(\log n)$.   □

Let us make some remarks about Theorem 4.15. First it is important to notice that the cost studied here is expected over all possible orders to insert the objects in the I-DAG, but there is no hypothesis on the new object, by opposition to Lemma 4.2 where all objects, including the last one, are supposed to satisfy the randomization hypothesis. Secondly, in many applications, it is possible to find all conflicts with a new object from a single one in time proportional to the actual number of conflicts (for example by the use of some neigborhood notions). The faster location may be used as a first step of the insertion of a new object in the I-DAG. A last remark concerns the worst case ; though we are interested in randomized complexities, the faster location has a worst case running time $O(n)$ which is better than the general location step.

4.2.2   Queries

In some applications, queries consist in finding the regions having zero width which are in conflict with a given element of the object universe : this is just a special instance of a location substep. In such cases, the I-DAG can be used and the randomized analysis of Theorem 4.3 holds, provided that the query object q together with the set of objects S introduced in the I-DAG satisfy the randomization hypothesis, i.e. the $(n+1)!$ permutations of $\{q\} \cup S$ are likely to occur.

In other applications, regions and queries are such that the answer to a query consists of exactly one region of $\mathcal{F}_0(S)$ for any set S. Such a query will be answered by a location substep that will traverse only one path from the root to a leaf. This yields the following strong variant of Lemma 4.2.

Theorem 4.16 *Assume that regions and queries are such that the answer to a query consists of exactly one region of $\mathcal{F}_0(S)$ for any set S. Then any query can be answered in $O(\log n)$ time provided that the n objects that were inserted in the I-DAG have been inserted in random order.*

Proof. The proof is similar to the proof of Theorem 4.15. □

4.3 Applications

We study several cases in which the update conditions are fulfilled : one algorithm for convex hulls (Section 4.3.1), arrangements (Section 4.3.2) and Voronoi diagrams of line segments (Section 4.3.3.2). We will also see cases where they are not fulfilled : another algorithm for convex hulls, and the case Voronoi diagrams (Section 4.3.3.1).

4.3.1 Convex hulls

We consider the geometric problem of computing the convex hull of a set of points. We present in this section two algorithms that are both on-line and whose expected performances are optimal in any dimension. For simplicity, we expose here firstly the two dimensional case. The extension to higher dimensional spaces is quite straightforward and will be shortly described next.

First algorithm

Objects are points of the plane. Regions are determined by three points. The region (p,q,r) associated with p, q and r consists of the union of two half-planes : one bounded by line (pq) and not containing r and the other bounded by line (qr) and not containing p (see Figure 4.1). An object is in conflict with a region if and only if it lies in the region.

Now let p, q and r be three points in S, the set of points already inserted. The region associated with these points has zero width if and only if p, q and r are three consecutive vertices of the convex hull. So computing the convex hull of S is equivalent to computing the zero width regions.

Let us now describe the algorithm. Suppose that the I-DAG has been constructed for the points in S and that we want to insert a new point m. The

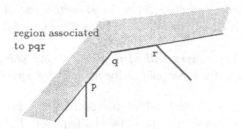

region associated
to pqr

q

r

p

Figure 4.1 : Definitions of regions for the convex hull problem

location substep gives the regions of $\mathcal{F}_0(\mathcal{S})$ containing $\mathfrak{m}$. If $\mathfrak{m}$ belongs to the interior of the convex hull, there is no such region, $\mathcal{F}_0(\mathcal{S} \cup \{\mathfrak{m}\}) = \mathcal{F}_0(\mathcal{S})$ and the I-DAG is not modified. Otherwise, let p_1 and p_k be the two vertices adjacent to $\mathfrak{m}$ in the new convex hull and $p_2, \ldots, p_{k-1}$ the chain of vertices which are no longer vertices of the convex hull after the insertion of $\mathfrak{m}$ (see Figure 4.2). The

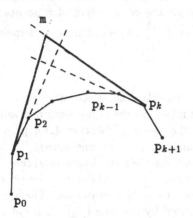

$\mathfrak{m}$

p_{k-1} p_k

p_2

p_{k+1}

p_1

p_0

Figure 4.2 : Inserting a new point in the convex hull

regions of $\mathcal{F}_0(\mathcal{S})$ containing $\mathfrak{m}$ are (p_{l-1}, p_l, p_{l+1}), for $1 \leq l \leq k$. By a simple test on these regions, we can determine (p_0, p_1, p_2) and (p_{k-1}, p_k, p_{k+1}) ($\mathfrak{m}$ belongs to only one of the two half-planes defining the region). The I-DAG is then modified in the following manner : all the (p_{l-1}, p_l, p_{l+1}), for $1 \leq l \leq k$, are killed by $\mathfrak{m}$, their width are incremented and three new regions are created, namely

- $(p_0, p_1, \mathfrak{m})$ as a son of (p_0, p_1, p_2)

- $(\mathfrak{m}, p_k, p_{k+1})$ as a son of (p_{k-1}, p_k, p_{k+1})

- $(p_1, \mathfrak{m}, p_k)$ as a son of both (p_0, p_1, p_2) and (p_{k-1}, p_k, p_{k+1}).

It is clear that the properties of the I-DAG are preserved and that the update conditions are satisfied. Here $f_0(r, \mathcal{S})$ is the expected size of the convex hull of r points of $\mathcal{S}$ which is clearly $O(r)$, so applying Theorem 4.3 we deduce :

Proposition 4.17 *The convex hull of n points in the plane can be computed on-line with $O(n)$ expected space and $O(\log n)$ expected update time.*

These results can be generalized to any dimension. The regions are determined by $d + 1$ points and are unions of two half-spaces. The zero width regions correspond to $(d-2)$-faces of the convex hull and their two adjacent $(d-1)$-faces. The expected size $f_0(r, \mathcal{S})$ of the convex hull of r points is $O\left(r^{\lfloor \frac{d}{2} \rfloor}\right)$.

Proposition 4.18 *The convex hull of n points in d-space can be computed on-line with $O\left(n^{\lfloor \frac{d}{2} \rfloor}\right)$ expected space and $O(\log n)$ expected update time if $d \leq 3$ and $O\left(n^{\lfloor \frac{d}{2} \rfloor - 1}\right)$ expected update time if $d > 3$.*

These results are optimal.

As far as queries are concerned, the above results show that one can decide if a point lies inside or outside the convex hull of n points in $O(\log n)$ expected time in the plane and $O\left(n^{\lfloor \frac{d}{2} \rfloor - 1}\right)$ expected time in d-space.

Second algorithm

It might look more natural to take half-spaces as regions. In that case, as will become clear below, the number of sons is not bounded and Update condition 1 is not satisfied. However the result of Section 4.1.4 proves that the resulting algorithm has the same complexity as the one above.

We describe the algorithm for the two dimensional case. It can be generalized to any dimension with no difficulty. $\mathcal{O}$ still denotes the set of points of the plane. Regions are now determined by only two points. The regions $(\mathbf{p}, \mathbf{q})$ and $(\mathbf{q}, \mathbf{p})$ are the two half planes limited by the line $(\mathbf{pq})$. A point is in conflict with the region $(\mathbf{p}, \mathbf{q})$ if it lies inside the corresponding half-plane. If a region $(\mathbf{p}, \mathbf{q})$ has zero width, then $[\mathbf{pq}]$ is an edge of the convex hull.

In addition to the standard information stored in each node of the I-DAG, we also maintain at each leaf which is associated with an edge E of the convex hull two pointers towards the two leaves associated with the two edges of the convex hull adjacent to E. When a new point $\mathbf{m}$ is inserted, the location substep provides all the half-planes with current width 0 in conflict with $\mathbf{m}$. These half-planes correspond to a chain of edges of the convex hull $[\mathbf{p}_1\mathbf{p}_2], \ldots, [\mathbf{p}_{k-1}\mathbf{p}_k]$ (see Figure 4.2). The two extremal edges $[\mathbf{p}_1\mathbf{p}_2]$ and $[\mathbf{p}_{k-1}\mathbf{p}_k]$ are identified, by testing if their two neighbors are not both in conflict with $\mathbf{m}$. Two new regions $(\mathbf{p}_1, \mathbf{m})$ and $(\mathbf{m}, \mathbf{p}_k)$ are created. In order to satisfy Property 2, $(\mathbf{p}_0, \mathbf{p}_1)$ and $(\mathbf{p}_1, \mathbf{p}_2)$ are made parents of $(\mathbf{p}_1, \mathbf{m})$; similarly $(\mathbf{p}_{k-1}, \mathbf{p}_k)$ and $(\mathbf{p}_k, \mathbf{p}_{k+1})$ are made parents of $(\mathbf{m}, \mathbf{p}_k)$. The neighborhood relationships are updated : $(\mathbf{p}_0, \mathbf{p}_1)$ and $(\mathbf{p}_1, \mathbf{m})$ become adjacent and similarly $(\mathbf{p}_k, \mathbf{p}_{k+1})$ and $(\mathbf{m}, \mathbf{p}_k)$, and $(\mathbf{p}_1, \mathbf{m})$ and $(\mathbf{m}, \mathbf{p}_k)$.

Notice than the width of (p_0, p_1) is still zero and that this region may have other sons in the future : Update condition 1 is not satisfied.

As described in Section 4.1.4, we introduce the notion of a bicycle. Here a bicycle is defined by two regions (p, q) and (q, r) sharing a point of definition (a bicycle here is a region of the previous section ; see Figure 4.1). The zero width bicycles are of two kinds. The first ones are associated with two regions with zero width : (p_0, p_1) and (p_1, m) in Figure 4.2. They correspond to two consecutive edges of the convex hull. The second ones are associated with a region of zero width and a region in conflict with the additional point of definition of the other : (p_1, p_2) and (p_1, m) in Figure 4.2. They correspond to an edge E of the convex hull and to an edge E', incident to one of the endpoint of E and whose supporting line separates a 1-set of S. Using the results on the number of k-sets (see, for instance, [Ede87]) we conclude that $g_0(r)$ is $O\left(r^{\lfloor \frac{d}{2} \rfloor}\right)$. Thus Theorem 4.13 implies that this simpler algorithm has the same complexity as the algorithm of the previous section.

Experimental results

The second algorithm has been coded in both dimensions 3 and 4. The results appear to be really good in the case where there are a lot of sites on the convex hull. In the case where the number of faces of the convex hull is small, compared to the number of sites, a lot of points must be located in the I-DAG, without creating any new result (this problem will not occur in the case of Voronoi diagrams, as will be seen next). Observe however that, in most cases, the number of nodes visited to locate a site is much smaller than the size of the current convex hull. Compare for example the results of Figures 4.3 and 4.4 in 3D.

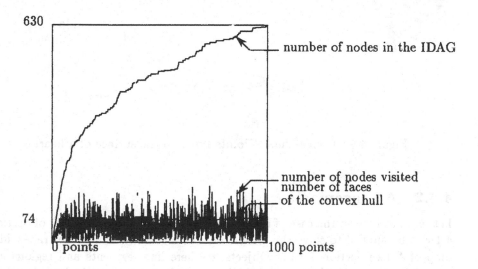

Figure 4.3 : Convex hull : 1000 sites in the interior of a 3D-cube

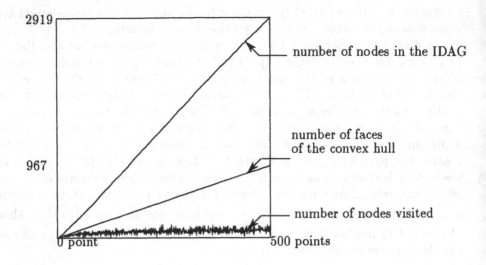

Figure 4.4 : Convex hull : 500 sites on the surface of a 3D-sphere

Figure 4.5 shows the results obtained for a set of points lying on a closed surface in 3D. Finally, Figures 4.6 and 4.7 show results in dimension 4.

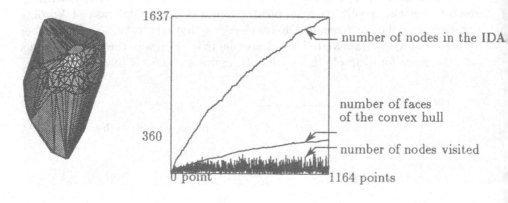

Figure 4.5 : Convex hull : Points lying on the surface of a heart

4.3.2 Arrangements

Let us consider first the case of line segments. The general framework of Section 4.1 can be applied to solve this problem. The algorithm builds the trapezoidal map of S (see Section 1.3.2). Objects are here line segments and regions are trapezoids (i.e. cells of the trapezoidal diagram). A trapezoid is determined by at most four segments. A line segment and a trapezoid are in conflict if and only if the segment intersects the interior of the trapezoid. For each leaf of the

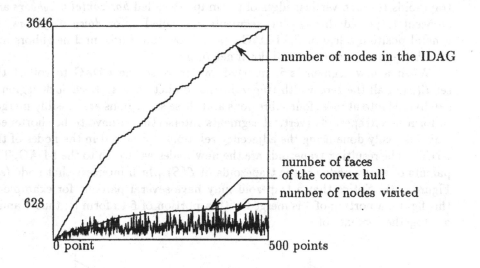

Figure 4.6 : Convex hull : Points lying in the interior of a 4 dimensional cube

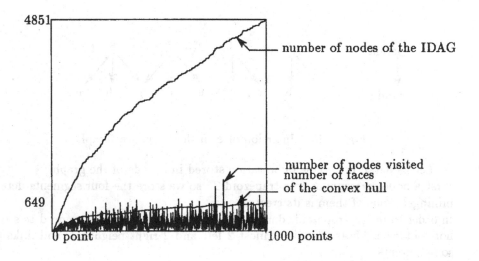

Figure 4.7 : Convex hull : Points lying on the surface of a 4 dimensional cube

I-DAG, we store also some neighbors of the corresponding zero width trapezoid. More precisely, we store two kinds of adjacency relationships : adjacent trapezoids through vertical edges of the map are called *horizontal neighbors* and adjacent trapezoid through line segments are called *up* or *down neighbors*. In general position a leaf of the I-DAG has at most four horizontal neighbors and an arbitrary number of up and down neighbors.

When a new segment **s** is inserted, we traverse the I-DAG to collect the set $\mathcal{L}(\mathbf{s})$ of all the zero width trapezoids in conflict with **s**. Each such region is subdivided into at most four subregions and these subregions are possibly merged to form new trapezoids (vertical segments intersecting **s** have to be shortened) which is easily done using the adjacency relationships stored in the nodes of the I-DAG. The resulting trapezoids are the new nodes we attach to the I-DAG. The parents of a new node are the trapezoids of $\mathcal{L}(S)$ which intersect that node (see Figure 4.8). Notice that a trapezoid may have several parents, for example in this figure, a portion of 4 is merged with a portion of 6 to form c ; thus 4 and 6 are together parents of c.

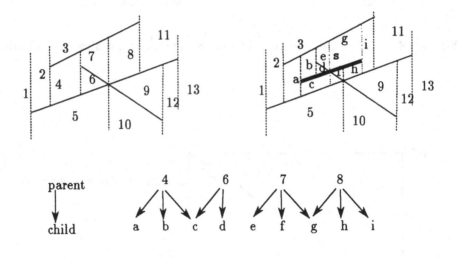

Figure 4.8 : Insertion of **s** in the Influence Graph

Let us detail the whole information stored in a node of the graph.

First a node corresponds to a trapezoid T, so we store the four segments determining T ; one of them is its creator.

In order to merge trapezoids during the insertion of a segment, we need to store horizontal neighbors, that is at most 2 left and 2 right neighbors, and links to some parents.

For the removal of a segment (see Section 6.2), we will need to store for each trapezoid some vertical neighbors : the up-right neighbor, namely its up neighbor adjacent to its right side neighbor, and the up-left, down-right and down-left neighbors, at the time when the trapezoid was created. To initialize these neighbors, for a new trapezoid, during an insertion, we need to know all current up and down neighbors of its parents. We have two ways to perform this : the first one is to maintain all those vertical neighbors for each trapezoid in the map ; the second one is to look for them each time when we need them ; to do so, we only need one of them, and find the other ones using horizontal neighbors. The first solution makes the structure of a node rather heavy, and we can easily see that the second solution does not increase the time complexity and gives all necessary information [CS89]. So we will consider that we know every up and down neighbor.

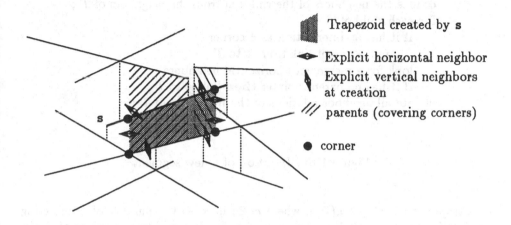

Trapezoid created by **s**

Explicit horizontal neighbor

Explicit vertical neighbors at creation

parents (covering corners)

corner

Figure 4.9 : A trapezoid, its neighbors and parents

Let us summarize the structure of a node (see Figure 4.9) :
- at most 4 segments defining it (one of them is the creator)
- the at most 4 horizontal current neighbors
- its at most 4 children
- its killer
- its at most 4 corners (i.e. its vertices)
- all current vertical neighbors (they are not explicitly stored)
- one vertical neighbor at creation per corner
- at most 4 parents : for each corner, the parent covering it

If the node is dead, all current neighbors become neighbors at the death.

Figure 4.10 shows the insertion of a new segment **s**. **s** is the new segment and T is a node of the Influence Graph. For the first call to Insert, T is the root of the Influence Graph.

A node has at most four sons so that the update conditions are fulfilled (but the number of parents for a trapezoid is unbounded). An easy lemma, proved in [CS89], shows that $f_0(r, \mathcal{S}) = O\left(r + a\frac{r^2}{n^2}\right)$. In fact, if $\mathcal{R}$ is a random subset of $\mathcal{S}$ of size r, since the trapezoidal map of $\mathcal{R}$ is a planar map, the number of

Insert(**s**, T)
    if s is not in conflict with T
        return ;
    if T is dead **then**
        for each child S of T Insert(**s**, S) ;
    else
        s is the killer of T ;
        split T into pieces, that become children of T ;
        deduce the neighbors of the children from the neighbors of T ;
        for each child of T
            if it has an intersection as a corner
                put a parent link from it to T ;
            look for a merge with horizontal neighbors ;
            if it has no parent pointer **then** link it to T ;
        update all neighbor relations of the neighbors of T ;

Figure 4.10 : Insertion of a new segment

its trapezoids is $O(n + a(\mathcal{R}))$, where $a(\mathcal{R})$ denotes the number of intersecting pairs of segments of $\mathcal{R}$. Let **z** be an intersection between two segments of $\mathcal{S}$, the probability that **z** be an intersection between segments of $\mathcal{R}$ is $\frac{\binom{r}{2}}{\binom{n}{2}}$, since the two segments meeting at **z** must both be in $\mathcal{R}$. The result follows from summing this probability for all intersections between segments of $\mathcal{S}$.

This result can be readily extended to planar arrangements of curves of bounded degree : the expression for $f_0(r, \mathcal{S})$ is the same, and the number of sons of a node is bounded by a constant depending on the kind of curves considered (it is 11 for circles, for example).

Proposition 4.19 *An arrangement of n planar curves (of bounded degree) can be computed on-line with $O(n + a)$ expected space and $O\left(\log n + \frac{a}{n}\right)$ expected update time, where a is the complexity of the arrangement.*

We achieve the same complexity as Clarkson and Shor [CS89] and Mulmuley [Mul89b], whose algorithms use the conflict graph, and thus are static. The trapezoids with zero width partition the plane, so that Theorem 4.16 applies :

Proposition 4.20 *A point can be located in an arrangement of n planar curves (of bounded degree) in $O(\log n)$ expected time using $O(n + a)$ expected space and $O(n \log n + a)$ expected preprocessing time, where a is the complexity of the arrangement.*

4.3.3   Voronoi diagrams

4.3.3.1   Voronoi diagrams of point sites in $\mathbb{E}^d$

Using the well known correspondence between Voronoi diagrams in d dimensions and convex hulls in $d + 1$ dimensions (see Section 1.4), we immediately deduce from the previous section two optimal on-line algorithms to construct the Voronoi diagrams of points in any dimension.

A direct presentation that does not use this correspondence has already been described in detail in Chapter 3 : the Delaunay Tree is in fact nothing else but an Influence Graph. Here, Update condition 1 is not fulfilled, so, for the analysis, we need to define a bicycle as a pair of simplices sharing a facet. We use the fact that the number of simplices arising in a Delaunay triangulation of n sites is $O\left(n^{\lfloor \frac{d+1}{2} \rfloor}\right)$ (see for example [Kle80]) and thus, that the number of bicycles of width zero has the same complexity (d is fixed). The analysis given in Theorem 4.13 allows to state :

Proposition 4.21 *The Delaunay triangulation of n points in d-space can be computed on-line with* $O\left(n^{\lfloor \frac{d+1}{2} \rfloor}\right)$ *expected space in dimension $d \geq 2$. The expected update time is $O(\log n)$ if $d = 2$ and $O\left(n^{\lfloor \frac{d+1}{2} \rfloor - 1}\right)$ if $d \geq 3$* [1].

These results are optimal.

> *Remark 4.22* The results stated in Proposition 4.21 are obtained by using worst-case bounds for the size of a Delaunay triangulation ($O\left(n^{\lfloor \frac{d+1}{2} \rfloor}\right)$ is used to bound $|\mathcal{F}_0(\mathcal{S})|$).
>
> If we assume that the set of sites to be triangulated is uniformly distributed in the space, we know from the results of [Dwy91] that $|\mathcal{F}_0(\mathcal{S})|$ (and thus $|\mathcal{G}_0(\mathcal{S})|$) is $O(n)$. Consequently, the random sampling technique used in Theorem 2.3 provides a linear complexity for the size of $\mathcal{F}_{\leq j}(\mathcal{S})$ and $\mathcal{G}_{\leq j}(\mathcal{S})$ and our algorithm runs in $O(n \log n)$ expected time in any dimension.

Experimental results

The algorithm has been implemented in both dimensions 2 and 3. It is to be noted that the algorithm is extremely simple.

Moreover, the numerical computations involved are also very simple : when a simplex is created, we can compute the coordinates of the center of its circumscribing sphere, and its squared radius. This is achieved by first writing the equation of a sphere in d dimensions, then writing that the $d + 1$ sites defining the simplex belong to the sphere, and solving the linear system which results

---

[1] The constants depend on the dimension as a $\frac{(d+5)!}{\lceil \frac{d}{2} \rceil!}$ factor for the time complexity and a $\frac{1}{(\lceil \frac{d}{2} \rceil - 1)!}$ factor for the space complexity. We will omit the computations of these constants here.

from that, with $d+1$ equations and $d+1$ unknowns, in $O(d^3)$ time.

The center and the squared radius of a simplex is computed once for all and stored in the corresponding node. For testing if a site is in conflict with a simplex, we only have to compute its squared distance to the center of the simplex, which costs $O(d)$, and compare it with the squared radius. The constant in Footnote 1 has been computed according to this method.

The algorithm has run on many examples with different kinds of point distributions, in two and three dimensions.

In each case, several orders of insertion of the sites have been tried to analyze the running time down to the constants. All randomized orders that we experimented yielded roughly the same results.

Some results are presented on Figures 4.11, 4.12 and 4.13 for the planar case. See also some related figures in Chapter 5 (5.8 to 5.16) and in Chapter 6 (6.9 to 6.11).

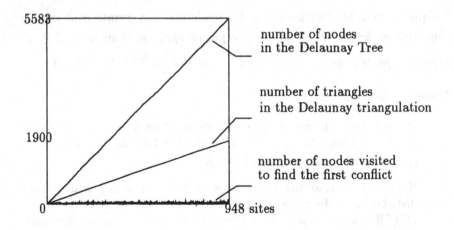

Figure 4.11 : Voronoi diagram : Random sites in the plane

Figures 4.14, 4.15 and 4.16 present results in 3-space. Figures 4.14 and 4.15 show results in the case where the size of the final Delaunay triangulation is linear in the number of sites, whereas Figure 4.16 was obtained for a distribution where the number of tetrahedra was quadratic (the sites are lying on two non-coplanar line segments).

Storage of the Delaunay Tree

The number of nodes in the Delaunay Tree has been studied with respect to the number of inserted sites. This must be compared with the number of simplices in the final triangulation, which is the size of the output.

In the two dimensional case, the ratio between the number of nodes in the Delaunay Tree to the size of the output is less than 3, in any example. In the space, this ratio becomes 4 in the case where the size of the triangulation is linear, and only 2 in the quadratic case.

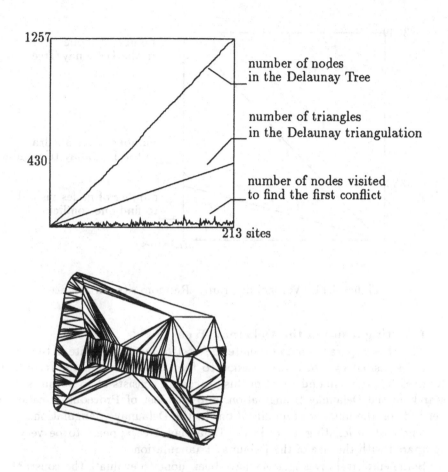

Figure 4.12 : Voronoi diagram : A non-convex curve

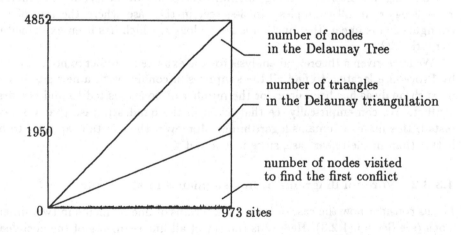

Figure 4.13 : Voronoi diagram : An ellipsis

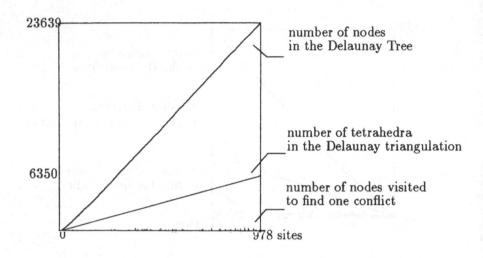

Figure 4.14 : Voronoi diagram : Random sites in 3-space

Inserting a site in the Delaunay Tree

The chosen parameter to evaluate the cost of inserting a site is the number of nodes visited by Procedure location to find the first conflict. According to Remark 3.3, the remainder cost of this procedure consists of an output sensitive search in the Delaunay triangulation, and the cost of Procedure creation also depends on the number of modifications in the Delaunay triangulation.

The cost of locating a site in the Delaunay Tree appears to be very small compared with the size of the Delaunay triangulation.

Some empirical investigations have been done to evaluate the constants. In the plane, the variations of the number of nodes visited to find the first conflict can be roughly (after smoothing) assimilated to the variations of the function $x \mapsto 3\log_2 x$, in all examples. In 3-space, in the case where the size of the triangulation is linear, the function is $x \mapsto 7\log_2 x$, which has been explained in Remark 4.22.

We have given a theoretical analysis for the expected number of nodes visited by Procedure location to find all the simplices in conflict with a new site **m**, but no such analysis has been done for the number of nodes visited to find the first conflict. We can empirically see that, even in the quadratic case that we have tested, this number remains logarithmic. Moreover, the constant appears to be better than in the linear case, since it is about 4.

4.3.3.2 Voronoi diagrams of line segments in the plane

Let us consider now the case of Voronoi diagrams of line segments in two dimensions (see Section 1.2.3). Here $\mathcal{O}$ is the set of all line segments of the euclidean plane. Let **p**, **q**, **r** and **s** be four segments and let Γ be the portion of the bisector of **p** and **q** extending between the two points equidistant from **p**, **q**, **r** and

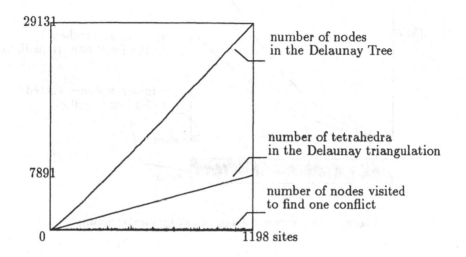

Figure 4.15 : Voronoi diagram : A closed surface (a heart)

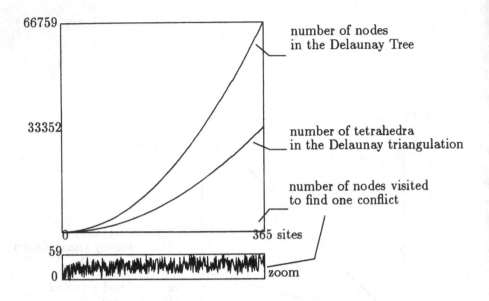

Figure 4.16 : Voronoi diagram : A quadratic example

p, q, s respectively (see Section 1.2.3). The region, noted (pq, r, s), associated with p, q, r and s is the union of the interiors of the disks tangent to p and q whose centers lie on Γ. Notice that, with this definition, a region is what was previously called a bicycle for the analysis of the Delaunay Tree. A line segment and a region are in conflict if and only if they intersect. A region has zero width if and only if Γ is an edge of the Voronoi diagram.

The update algorithm is as follows. We find in the I-DAG all regions in conflict with the new segment m : these regions correspond to the edges which disappear in the new diagram. The set of disappearing edges form a tree. In fact, an edge (or a part of an edge) E disappears if the points lying on it are closer to m than to the segments determining E. Then, after the insertion of m, these points will belong to the Voronoi cell $V(m)$ of m. Assume now that the set of disappearing edges contains a cycle. Necessarily, this cycle must surround at least one segment s. In the new diagram, the cell $V(s)$ will form a hole in $V(m)$, which is impossible because all cells of the diagram are simply connected.

Let E be a disappearing edge. If one of the two end-points of E is still a valid vertex of the Voronoi diagram (that is if the segment m does not intersect the corresponding disk) we compute in constant time the portion of E that remains in the new diagram. The new region associated with that new edge becomes a son of the region associated with E. We then connect the new vertices of the Voronoi diagram (which are new end-points lying on old edges) by edges supported by new bisectors (see Figure 4.17). The region corresponding to a new edge E' is made son of the regions associated with the unique path of disappearing edges that joins the two end points of E'. This ensures that this region is contained in its parents : an edge E' is the bisector of m and some $s \in S$. E' lies entirely in the Voronoi cell $V(s)$ of the diagram before the insertion of m. Let p be a point belonging to the region associated to E'. p lies in the interior of a disk tangent to both m and s, which has no conflict. In the previous diagram, before the insertion of m, this disk can grow larger, without having any conflict, until it touches one segment of S. Then its center necessarily lies on some edge on the boundary of the previous cell $V(s)$. So p lies in the region associated to one father of the region associated to E'.

The update conditions are satisfied and, by the Euler relation, f_0 is linearly related to the number of segments.

> *Remark 4.23* We can remark that, if the analogous definition had been given for regions in the case of point sites only, we would have obtained another possible definition for the Delaunay Tree, that would have satisfied Update condition 1.

Proposition 4.24 *The Voronoi diagram of n line segments in the plane can be computed with $O(n)$ expected space and $O(\log n)$ expected update time.*

Notice that the algorithm has been coded.

4.4   About complexity results

4.4.1   Randomization

Our analysis of the space and time required to build the I-DAG structure is randomized. As previously noted, randomization concerns here only the order in which the inserted objects are introduced in the structure. No assumption is made as to the distribution of the input. As already said, our results are expected values that correspond to averaging over the $n!$ possible permutations of the n inserted objects, each supposed to be equally likely to occur.

4.4.2   Amortization

We have been able to bound the cost of inserting the kth object in the I-DAG. This cost is not amortized as opposed to the results in [BDT93, GKS92] but the kth object may be any one of the inserted objects with the same probability.

It must be noted however that the bound given in Theorem 4.3 cannot be a bound for the cost of inserting a given object. Indeed let us consider the construction of the Delaunay triangulation (the dual of the Voronoi diagram) of a set of n points in the plane. We take $\frac{n-1}{2}$ points close to one line segment, $\frac{n-1}{2}$ points close to another line segment, and one point, say o, between the two segments (see Figure 4.18). For appropriate positions of the points, the insertion of o will modify most of the triangles; thus the expected cost of inserting o in the I-DAG at step k is $\Omega(k)$ whatever k is.

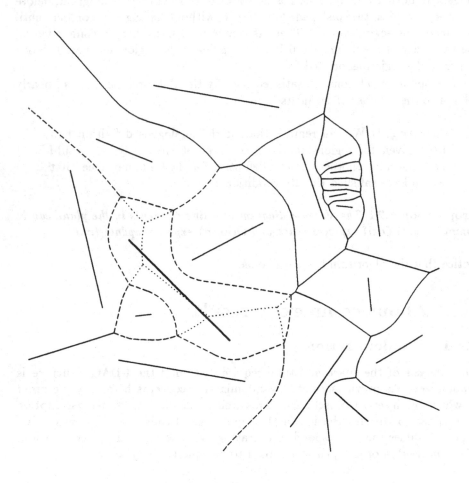

The new segment **m** is in bold line.
The dotted edges correspond to regions in conflict with **m**.
The dashed edges correspond to new regions created by **m**.

Figure 4.17 : Insertion of **m** in the Voronoi diagram

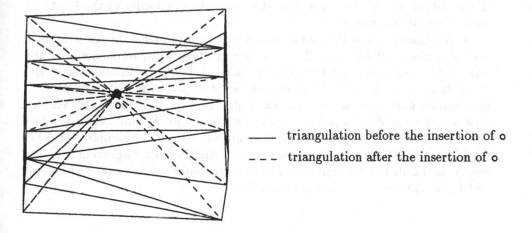

—— triangulation before the insertion of o

‐ ‐ ‐ triangulation after the insertion of o

Figure 4.18 : Cost of inserting point o

This is not in contradiction with our result. Indeed, the cost of inserting a given object appears weighted by the probability factor $1/n$ in the expected cost of step k. Our bound on the cost of the kth insertion proves that objects requiring expensive updates are rare whatever the set S of input data may be.

4.4.3   Output sensitivity

An algorithm is said to be output-sensitive if, for a given set of input data, its complexity depends on the actual size of the output. It is clearly impossible, in general, to have incremental algorithms that are sensitive to the final output because at some stage of the incremental construction the intermediate results may be greater than the final one. We may illustrate this with the example of the Voronoi diagram in three dimensions. Let S be a set of n points lying on two non coplanar line segments. The Voronoi diagram of S is quadratic but, if a point o between the two segments is added to S, the diagram of $S \cup \{o\}$ becomes linear.

In view of this fact, it is interesting to define *on-line output sensitive* algorithms as algorithms whose update complexities depend on the actual size of the current output.

Our algorithms are not on-line output sensitive because the expected complexity of each step depends on $f_0(r, S)$ (or $g_0(r, S)$) for some $r \leq n$. Let us consider again the case of the Voronoi diagram of the set of points above. Inserting a $n + 2^{nd}$ point to $S \cup \{o\}$ will take $O(n^2)$ expected time using the I-DAG although the current output is $O(n)$.

However, in many situations, the expected value $f_0(r, S)$ is a well behaved function of the size r of the random sample which is sensitive to the actual size of the output for S. In such a case, the expected complexity of the I-DAG is on-line output sensitive.

A first illustration is the case of an arrangement of planar curves which has been described in Section 4.3.2. As a second illustration, let us consider the case of the Voronoi diagram in higher dimensions. For some distributions of the input data, the diagrams built on the entire set of points as well as on most of the samples have a linear size. For example, the expected size of the Voronoi diagram of a set S of n points evenly distributed in the unit d-ball is $O(n)$ and the expected size of the Voronoi diagram for a r-random sample $f_0(r, S)$ is $O(r)$ [Dwy91]. This result readily implies that the Voronoi diagram of n points evenly distributed in the unit d-ball can be computed on-line with $O(n)$ space and $O(\log n)$ update time in any dimension.

Conclusion

We have presented in this chapter a general framework for the design and analysis of on-line algorithms. This framework has been applied successfully to various problems : convex hulls and Voronoi diagrams in any dimension, Voronoi diagrams of segments in the plane, arrangements of curves in the plane. The algorithms are randomized, simple and in some cases output sensitive. They have been coded easily and preliminary experiments have provided strong evidence that they are very efficient in practice. The I-DAG can be used to solve several other problems and provides simple on-line algorithms with the same worst-case complexities as the best (in general static) deterministic algorithms. We simply mention some of them :

- Computing the upper envelope of triangular surface patches in the three dimensional space, with a complexity of $O\left(n^2\alpha(n)\log n\right)$ in the worst case where the size of the output is $O\left(n^2\alpha(n)\right)$ [BD92].

- Computing abstract Voronoi diagrams ([MMO91] propose a static randomized algorithm).

- Computing arrangements of surface patches in space (see [CEGSW90] for combinatorial bounds).

- Computing the intersection of n half spaces : this problem is dual to constructing convex hulls.

- Computing the union of n balls in d space : consider the d-dimensional space as an hyperplane of a $d+1$-space and use an inversion with a point outside the hyperplane as its pole : the problem reduces to that of computing the intersection of n half spaces.

- Computing the visibility graph of a set of line segments in the plane (see [Wel85]) : take as regions the triangles containing two edges of the visibility graph incident to a common vertex and consecutive when sorted by polar angle around this vertex. A line segment is in conflict with a region if it intersects the region.

Our technique assumes that the geometric structure to be computed is closely related to the regions of zero width. One may be interested in computing instead regions of width $\leq k$ to construct, for example, k-sets or Voronoi diagrams of order k. It is possible to generalize the I-DAG and to obtain results similar to the ones described here. The complexity results will depend on the expected number of regions of width $\leq k$ defined by random samples, as will be seen in Chapter 5.

We will study the possibility of allowing deletions as well as insertions, thus making the structure fully dynamic, in Chapter 6.

Chapter 5
The *k*-Delaunay Tree

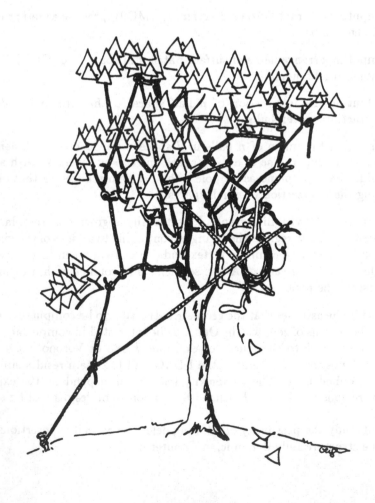

The k-Delaunay Tree is an extension of the Delaunay Tree, and is used for the construction of higher order Voronoi diagrams defined in Section 1.2.2.

K.L. Clarkson's randomized algorithm [Cla87] determines the order k Voronoi diagram of n sites in the plane in expected time $O(kn^{1+\epsilon})$ with a constant factor that depends on ϵ.

K. Mulmuley [Mul91a] achieves a complexity of $O\left(k^{\lceil\frac{d+1}{2}\rceil}n^{\lfloor\frac{d+1}{2}\rfloor}\right)$ for $d \geq 3$ ($O(nk^2 + n\log n)$ if $d = 2$), for the randomized construction of the Voronoi diagram of order 1 to k (see Section 2.3.2). These algorithms are static.

We present here an algorithm that is semi-dynamic. After each insertion of a new site, the algorithm updates a data structure, called the k-Delaunay Tree [BDT93]. This structure generalizes the Delaunay Tree to compute the Delaunay triangulation (and, by duality, the Voronoi diagram) of a set of points. The k-Delaunay Tree contains all the successive versions of the order $\leq k$ Voronoi diagrams and allows fast point location.

As in the preceding chapter, we show that randomization allows to obtain an efficient complexity on the average, which is, as usual, impossible in the worst case for a semi-dynamic algorithm. Our main result states that if we randomize the insertion sequence of the n sites, the k-Delaunay Tree (and thus the order $\leq k$ Voronoi diagrams) can be constructed in expected time $O(n\log n + k^3 n)$ in two dimensions and expected storage $O(k^2 n)$.

Our algorithm extends to higher dimensions. For a given value d of the dimension, its expected time complexity is $O\left(k^{\lceil\frac{d+1}{2}\rceil+1}n^{\lfloor\frac{d+1}{2}\rfloor}\right)$ and its expected space complexity is $O\left(k^{\lceil\frac{d+1}{2}\rceil}n^{\lfloor\frac{d+1}{2}\rfloor}\right)$.

Very recently, F. Aurenhammer and O. Schwarzkopf [AS92] proposed a dynamic algorithm for the construction of the order k diagram in the plane whose randomized complexity is $O\left(k^2 n + kn\log^2 n\right)$ (see Chapter 7).

The overall organization of the chapter is the following. In Section 5.1, we define the k-Delaunay Tree in two dimensions and present an algorithm for its construction. In Section 5.2, we analyse the complexity of the randomized construction of the k-Delaunay Tree and thus, of all order $\leq k$ Voronoi diagrams. Section 5.3 shows how to use the k-Delaunay Tree for searching the l nearest neighbors of a given point. In Section 5.4, we extend our results to d dimensions. Last but not least, Section 5.5 presents experimental results which provide evidence that the algorithm is very effective in practice for small values of k.

5.1 The k-Delaunay Tree in two dimensions

$\mathcal{S}$ is a non degenerated set of n sites in the euclidean plane. We use the terms defined in Section 1.2.2.

We first show a lemma which will be useful in the sequel.

Lemma 5.1 Let $\mathcal{T} \subset \mathcal{S}$. The Voronoi polygon $V(\mathcal{T})$ does not change if we add to $\mathcal{T}$ and $\mathcal{S}$ some new points lying in the convex hull of $\mathcal{T}$. More precisely $V(\mathcal{T})$

in $Vor_{|T|}(S)$ is equal to $V(T \cup R)$ in $Vor_{|T \cup R|}(S \cup R)$ if the points of R lie in the interior of the convex hull of T.

Proof. Let R be a set of points not in S, and lying in the interior of the convex hull of $T \subset S$. We denote $V_S(T)$ the Voronoi polygon of T, where T is considered as a subset of S (i.e. in $Vor_{|T|}(S)$). We first prove that $V_S(T) \subset V_{S \cup R}(T \cup R)$.

Let $m \in V_S(T)$, $p \in T \cup R$, and $q \in (S \cup R) \setminus (T \cup R) = S \setminus T$.

- if $p \in T$, then $\delta(m, p) < \delta(m, q)$, since $m \in V_S(T)$.
- if $p \in R$, then $p = \sum_i \alpha_i t_i$, where $t_i \in T$ and $\alpha_i \in \mathbb{R}^+, \sum_i \alpha_i = 1$.

$$
\begin{aligned}
\delta(m, p) \quad &\leq \quad \sum_i \alpha_i \delta(m, t_i) \\
&\text{since } \delta \text{ is associated to the euclidean norm,} \\
&\text{and thus } x \mapsto \delta(m, x) \text{ is a convex function} \\
&< \quad \sum_i \alpha_i \delta(m, q) \text{ since } t_i \in T \\
&= \quad \delta(m, q)
\end{aligned}
$$

In both cases, $\delta(m, p) < \delta(m, q)$, from which we deduce that : $m \in V_{S \cup R}(T \cup R)$.

The reciprocal inclusion is straightforward, which achieves the proof. $\square$

5.1.1   Including and excluding neighbors

In the sequel, we will denote $B(T)$ the interior of the disk circumscribing triangle T.

For a triangle T, we will define 2 neighbors through each of its 3 edges : one will be called the *including* neighbor and the other one the *excluding* neighbor. This notion of neighborhood corresponds to the actual notion of adjacency in the higher order Voronoi diagrams.

Let E be an edge of T and p be the opposite vertex of T. Let us consider a moving disk B whose boundary passes through the end points of E, and whose center moves along the bisecting line of E. Starting from $B = B(T)$, we can move B in two opposite directions : the one such that $p \in B$ is called the including direction and the other such that $p \notin B$ is called the excluding direction. We stop moving B as soon as its boundary encounters a site different from the end points of E. Let q_i (resp. q_e) be the first site encountered in the including (resp. excluding) direction. The triangle T_i (resp. T_e) having E as an edge and q_i (resp. q_e) as a vertex will be called the *including neighbor* (resp. *excluding neighbor*) of T through edge E (see Figure 5.1). Notice that q_i and q_e may be on either side of the line supporting E.

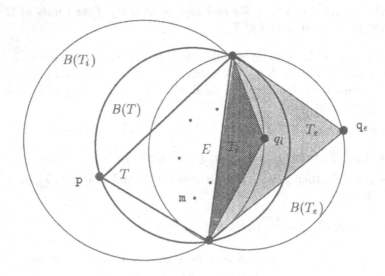

Figure 5.1 : Including and excluding neighbors

Remark 5.2 In the sequel, we will also speak of the neighbors of a triangle T in the direction including a site m in $B(T)$ and in the direction excluding m. The neighbor of T through E in the direction excluding m (resp. including m) is the excluding (resp. including) neighbor of T if m and p are on the same side of E and the including (resp. excluding) neighbor of T otherwise.

Remark 5.3 If S is the excluding neighbor through E for T, then T is a neighbor of S, but T may be either the including neighbor through E of S, if q_e lies on the same side as T with respect to the supporting line of E, or the excluding one on the other case.

In the same way, if S is the including neighbor of T, T is the including neighbor of S if q_i does not lie on the same side as T with respect to E, and the excluding neighbor otherwise.

The neighborhood relationships are reciprocal if and only if S and T do not lie on the same side with respect to the supporting line of E. They are inversed otherwise

Remark 5.4 The following property will be useful in the sequel :

$$B(T) \subset B(T_i) \cup B(T_e)$$

Hence, if a site m lies inside $B(T)$, we can deduce that m lies into either $B(T_i)$ or $B(T_e)$.

If $B(T)$ contains k sites, $B(T_e)$ contains k sites if q_e lies outside $B(T)$ (the k sites in $B(T)$), or $k-1$ sites if q_e lies inside $B(T)$ (the k sites in $B(T)$ minus

q_e). Speaking in terms of Voronoi vertices, if $B(T)$ contains k sites, T is dual to a close-type vertex τ of Vor_{k+1} and its excluding neighbors are dual to the adjacent vertices of τ in Vor_{k+1}.

Similarly, if $B(T)$ contains k sites, then $B(T_i)$ contains $k + 1$ sites if q_i lies outside $B(T)$ (the k sites in $B(T)$ plus p), or k sites if q_i lies inside $B(T)$ (the k sites in $B(T)$ plus p minus q_i). Then T is dual to a far-type vertex τ' of Vor_{k+2}, and its including neighbors are dual to the adjacent vertices of τ' in Vor_{k+2}.

5.1.2   A semi-dynamic algorithm for constructing the order $\leq k$ Voronoi diagrams

Our algorithm is a generalization of the semi-dynamic algorithm presented in Chapter 3 for constructing the Delaunay triangulation. Each site is introduced, one after another, in each of the order $\leq k$ Voronoi diagrams and each diagram is subsequently updated. We will describe this algorithm at the same time as the k-Delaunay Tree, in the following sections, but it is actually independent of that data structure.

5.1.3   Construction of the k-Delaunay Tree

The k-Delaunay Tree is a hierarchical structure that we use to construct the order $\leq k$ Voronoi diagrams. As in the Delaunay Tree, the nodes are associated with triangles. We will often use the same word *triangle* for both a triangle and the node associated with it. The k-Delaunay Tree is not really a tree but a rooted direct acyclic graph.

Recall that the width of a triangle is the number of sites in conflict with it or, in other words, the number of sites lying in the interior of its circumscribing disk.

Following our algorithm, each site is introduced in turn and we keep all triangles in a hierarchical manner in the k-Delaunay Tree, creating appropriate links between "old" (i.e. created before the introduction of the site) and "new" triangles (i.e. created after the introduction of the site). This will allow to efficiently locate a new site in the current structure.

5.1.3.1   Initialization

For the initialization step we choose 3 sites. They define one finite triangle and six half planes (considered as infinite triangles) limited by the supporting lines of the finite triangle. Three of the half planes contain one site, the other ones are empty. The 4 triangles of current width zero will be the sons of the root of the tree. Their neighborhood relationships in the Delaunay triangulation are created. The 3 triangles of current width 1 are linked to the preceding ones by their neighborhood relationships in the order 2 Voronoi diagram.

5.1.3.2   Inserting a new site

The k-Delaunay Tree will be constructed so that it satisfies the following property :

> ($\mathcal{P}$) *all the triangles of current width strictly less than* k *are present in the* k-*Delaunay Tree.*

Hence, the triangles dual to the vertices of the order $\leq k$ Voronoi diagrams are all present in the structure. Moreover, we keep their adjacency relationships in the corresponding Voronoi diagrams. The k-Delaunay Tree thus contains the whole information necessary to construct all the order $\leq k$ Voronoi diagrams.

Let us suppose that the k-Delaunay Tree has been constructed for the already inserted sites and satisfies the above property ($\mathcal{P}$). Let m be the next site to be inserted. We will see how to update the k-Delaunay Tree so that it still satisfies ($\mathcal{P}$) after the insertion of m.

Let S be a triangle dual to a vertex of some of the order $\leq k$ Voronoi diagrams, to be created after the insertion of m in order to satisfy ($\mathcal{P}$). S is called a *new triangle*. S has m as a vertex ; let E be the edge of S not containing m. S has an including neighbor T through E, and an excluding neighbor R through E. It is plain to observe that R and T were necessarily neighbors before the insertion of m and that $B(T)$ contains m while $B(R)$ does not. We have obtained a necessary condition for the creation of a triangle. Figure 5.2 shows the four typical situations.

Conversely, we show that this condition is sufficient. Let T be a triangle dual to a vertex of some of the order $\leq k$ Voronoi diagrams, whose disk $B(T)$ contains m. Let E be an edge of T, and p be the third vertex of T. If the neighbor R of T through E in the direction excluding m does not contain m in its disk, then T and R cannot stay neighbors any longer. We must create a new triangle S by linking m to E. T becomes the including neighbor of S through E and R its excluding one. (Notice that, in Case (d) of Figure 5.2, if $l = k$, R does not need any including neighbor, which would belong to the Voronoi diagram of order $k + 1$, so S is not created in this case.)

If now the current width of S is l, the current width of T, before the insertion of m, was l or $l - 1$ and the current width of R is still l or $l - 1$, as before the insertion of m (see Figure 5.2). Thus property ($\mathcal{P}$) implies that T and R were associated to nodes in the k-Delaunay Tree before the insertion of m, the current width of T is increased by one and is now $l + 1$ or l. If this number is larger or equal to k, then T is marked as *dead* (T is no longer dual to a vertex of some order $\leq k$ Voronoi diagram). (Remember that, if $l = k$, which can happen only in Case (d), S is not created.)

Furthermore, if the current width of S is zero when it is created (which can only occur in Case (a) of Figure 5.2), we then update the tree by making S a *son* of T and a *stepson* of R.

> *Remark 5.5* We can thus notice that the subgraph of the k-Delaunay
> Tree obtained by recursively traversing all links to sons and stepsons,

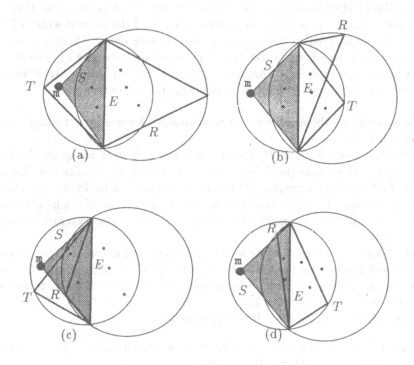

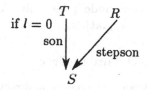

If l is the current width of the new triangle S, the current width of T before the insertion of $\mathbf{m}$, the current width of T after the insertion and the current width of R are respectively :

- $l, l + 1$ and l in the upper-left case (a)
- $l - 1, l$ and l in the upper-right case (b)
- $l, l + 1$ and $l - 1$ in the bottom-left case (c)
- $l - 1, l$ and $l - 1$ in the bottom-right case (d).

Figure 5.2 : Inserting a new site

starting from the root, is the Delaunay Tree presented in Chapter 3 (which is also the 1-Delaunay Tree) : a triangle can be created in the (1-)Delaunay Tree only if its current width is zero ; at this moment, it is the neighbor of its stepfather, and the current width of its father is incremented, so it becomes at once dead. There are no additional neighborhood relationships involving triangles of current width larger than zero, because the (1-)Delaunay Tree only deals with neighborhood relationships between Delaunay triangles.

We also update the neighborhood relationships between the triangles which are not marked as dead.

In order to ensure that $(\mathcal{P})$ still holds, it is sufficient to apply the above procedure to *all* the triangles T whose circumscribing disks contain **m**. Thus we need to find all those triangles. This will be performed by Procedure location while the creation of the new triangles and the maintenance of the neighborhood relationships will be performed by Procedure creation. These two procedures are detailed below.

When a node receives sons, its width increases, and it can only get new sons when its width is zero, so the total number of sons of a node is at most 3. Notice however that the number of stepsons is unbounded ; unfortunately, all stepsons are useful as shown in Chapter 3.

Nevertheless, we have the following property :

Lemma 5.6 *The total number of links to sons and stepsons in the k-Delaunay Tree is less than twice the number of nodes.*

Proof. Each newly created node (the 7 first ones excepted) has exactly one father and one stepfather. □

5.1.3.3 Structure of the k-Delaunay Tree

As previously noted, the k-Delaunay Tree is a directed acyclic graph. Each node of the k-Delaunay Tree is associated to a triangle. The node associated to triangle T contains the following informations :

- three links to the three sites which are vertices of T

- the center and the radius (more exactly the squared radius) of $B(T)$, which can also be used to mark T as finite or infinite

- its at most 3 sons

- the list of its stepsons

- the last site located in T

- the last site propagated in T (these last two fields will be used by Procedure location)

- the current width of T, which can also be used to mark T as dead

- the three excluding neighbors of T, and its three including neighbors if its width is at most $k - 2$

The first six fields are the same as in the (1-)Delaunay Tree (see Remark 5.5). In the (1-)Delaunay Tree, as soon as the current width of a triangle T is equal to 1, it becomes dead, so there is only a mark remembering whether T is dead or not ; another difference is that there are only three neighbors (they are excluding neighbors), that are the neighbors of T in the Delaunay triangulation, if T is not dead.

> *Remark 5.7* As already noticed, the two types of neighbors correspond to neighbors in different Voronoi diagrams, and allow, starting from one vertex of the order 1 Voronoi diagram, to reach all adjacent vertices in each of the higher order Voronoi diagrams. This point is crucial for Procedure **propagate**, called by procedure **location**, as will be seen in the next section.

5.1.3.4   Procedure location

If **m** belongs to the circumscribing disk of a triangle, we know (see Remark 5.4) that it belongs to the union of the circumscribing disks of its father and of its stepfather. So the traversal of the Delaunay Tree (Remark 5.5) gives all triangles whose disks contain **m**.

In fact, we traverse it until we find only one Delaunay triangle in conflict with **m**. Then we follow the neighborhood relationships to find all triangles, of current width strictly less than k, in conflict with **m** (see Remark 5.7). We store them in a list $T(\mathbf{m})$, the edges of these triangles are the possible candidates to be used for the creation of new triangles.

Procedure **location** and Procedure **propagate** are described in Figure 5.3. Observe that Procedure **location** is the same as in Chapter 3, because it consists of a traversal of the Delaunay Tree.

5.1.3.5   Procedure creation

We go through the list $T(\mathbf{m})$. Let T be a triangle of this list and E one of its edges. If the neighbor R of T through E in the direction excluding **m** does not contain **m**, we then create S by linking **m** to E, and the son and stepson relations involving S, if the current width of S is 0. Procedure **creation** is described in Figure 5.4.

Let us now see how we can maintain the neighborhood relationships between triangles.

As already mentioned, when S is created, R is the excluding neighbor of S through edge E common to T and R, and T is the including one. We can easily compute the reciprocal relations : R and T were neighbors before the insertion of **m** (notice that R may be either an excluding or an including neighbor of

Initialize the list $T(\mathbf{m})$ as the empty list
location($\mathbf{m}$,root of the k-Delaunay Tree)

Procedure location($\mathbf{m}$,node)
    ($\star$ The node is associated to triangle T $\star$)
        if the last site located in T is not $\mathbf{m}$ and
        if $\mathbf{m}$ lies into $B(T)$, **then**
            $\mathbf{m}$ becomes the last site located in T;
            for each son, location($\mathbf{m}$,son);
            for each stepson, location($\mathbf{m}$,stepson);
            if the current width of T is 0, **then**
                propagate($\mathbf{m}$,T);
                stop Procedure location.

Procedure propagate($\mathbf{m}$,T)
    $\mathbf{m}$ becomes the last site propagated in T;
    add T to the list $T(\mathbf{m})$;
    increment the current width of T;
    if the current width of T is now k, **then** mark T as dead;
    for each neighbor N of T
        if the last site propagated in N is not $\mathbf{m}$ and
        if N is not dead, **then**
            if $\mathbf{m}$ lies into $B(N)$, **then** propagate($\mathbf{m}$,N).

Figure 5.3 : Locating a site in the k-Delaunay Tree

Procedure creation($T(\mathbf{m})$)

    for each triangle T of $T(\mathbf{m})$
        for each edge E of T
        if the neighbor R of T through E in the direction
            excluding $\mathbf{m}$ does not contain $\mathbf{m}$ in its disk, **then**
          • create the triangle S having vertex $\mathbf{m}$ and edge E,
            and its associated node
            (except in the case where the width of S is k);
          • declare R as the excluding neighbor of S through E
            and update the reciprocal relation;
          • declare T as the including neighbor of S through E
            and update the reciprocal relation;
          • **if** the current width of S is 0, **then**
            create the relations : S son of T and S stepson of R;
    create the neighborhood relationships between the new triangles.

Figure 5.4 : Creating the new triangles

T ; remember also that the relations between neighbors are not symmetric, see
Remark 5.3). If T was the excluding (resp. including) neighbor of R through
E, then now S is the excluding (resp. including) neighbor of R. Similarly, if R
was the excluding (resp. including) neighbor of T through E, then now S is the
excluding (resp. including) neighbor of T.

In order to create the neighborhood relationships between the new triangles,
we proceed as follows. Let us suppose that the insertion of m creates $x(m)$ new
triangles $T_1, T_2, \ldots, T_{x(m)}$, where $T_i = (m, s_i^0, s_i^1), i = 1, 2, \ldots, x(m)$. The T_i's are
dual to vertices of various $Vor_l(\mathcal{S}), l \leq k$. The adjacency relationships must be
created in each $Vor_l(\mathcal{S}), l \leq k$. Our goal is to find, for each T_i $(i = 1, \ldots, x(m))$,
its excluding and its including neighbors through both edges ms_i^0 and ms_i^1. To
this aim, for each $s_i^j, i = 1, 2, \ldots, x(m), j = 0, 1$, we sort the list of new triangles
having s_i^j as a vertex, according to the abscissa of their center on the bisecting
line of $[m, s_i^j]$. By definition, (m, s_i^j, v) and (m, s_i^j, v') are neighbors through $[m, s_i^j]$
iff they are successive in this order. We thus only have to go through the
list of sorted triangles to obtain the neighborhood relationships through the
corresponding edge.

For each edge (m, s_i^j), the complexity of this sorting is $O(n_i^j \log n_i^j)$, where n_i^j
denotes the number of new triangles having (m, s_i^j) as an edge, which is bounded
by $2k$. The whole complexity of creating the neighborhood relationships is thus
$O(x(m) \log k)$ for each new site m.

> *Remark 5.8* The $\log k$ appearing in this complexity could be drop-
> ped, but it would complicate our algorithm. As we shall see in the
> sequel, the cost of Procedure creation is dominated by the cost of
> Procedure location, therefore we do not elaborate on improving its
> complexity.

5.2   Analysis of the randomized construction

The above algorithm allows to insert new sites in a dynamic way. As in Chapter
4, for the analysis, we assume that all sequences of insertion have the same
probability.

This section proves the main result of this paper, stated in the following
theorem :

Theorem 5.9 *For any set of n sites, if we randomize the sequence of their
insertion, the k-Delaunay Tree (and thus the order $\leq k$ Voronoi diagrams) of the
n sites can be constructed in expected time $O(n \log n + k^3 n)$ in two dimensions,
using expected storage $O(k^2 n)$.*

The remaining of Section 5.2 is devoted to the proof of Theorem 5.9. Section
5.2.2 analyzes the expected space used to store the k-Delaunay Tree, Section
5.2.3 the expected cost of locating the n sites and Section 5.2.4 the cost of
constructing the successive triangles and their neighborhood relationships.

5.2.1 Results on triangles and bicycles

Let $\mathbf{x}, \mathbf{y}, \mathbf{z}, \mathbf{t}$ be four sites. As in Chapter 4, the *bicycle* $\mathbf{x(yz)t}$ is the figure drawn by $B(\mathbf{xyz})$ and $B(\mathbf{yzt})$ (see Figure 5.5 which represents one of the possible configurations of a bicycle).

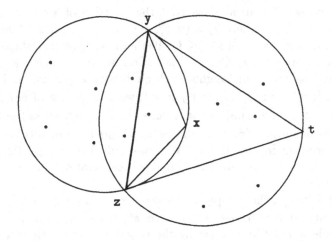

Figure 5.5 : The bicycle $\mathbf{x(yz)t}$

The width of a bicycle $\mathbf{x(yz)t}$ is the number of sites belonging to $B(\mathbf{xyz})$ and $B(\mathbf{yzt})$, where we do not take $\mathbf{x}, \mathbf{y}, \mathbf{z}$ and $\mathbf{t}$ into account. $\mathcal{F}_j(\mathcal{S})$ (resp. $\mathcal{G}_j(\mathcal{S})$) is the set of triangles (resp. bicycles) having width j and $\mathcal{F}_{\leq j}(\mathcal{S})$ (resp. $\mathcal{G}_{\leq j}(\mathcal{S})$) the set of triangles (resp. bicycles) of width at most j.

Lemma 5.10 *Let* $\mathbf{xyz}$ *be a triangle having width* j. $\mathbf{xyz}$ *will arise as a vertex of some order* $\leq k$ *Voronoi diagram during the construction with probability*

$$
\begin{cases}
\dfrac{k(k+1)(k+2)}{(j+1)(j+2)(j+3)} & \text{if } \ j \geq k \\
1 & \text{if } \ j < k
\end{cases}
$$

Proof. Let l be the current width of $\mathbf{xyz}$ when $\mathbf{xyz}$ is created. The property holds true if and only if one of the three sites $\mathbf{x}, \mathbf{y}$ and $\mathbf{z}$ is introduced after the l sites inside $B(\mathbf{xyz})$ (in any order, hence $3(l+2)!$ possible orders for the $l+3$ first sites, and $\dbinom{j}{l}$ possible sets of l sites among the j ones), and before the $j-l$ remaining sites (which may also be introduced in any order, that is $(j-l)!$ possibilities). There are $(j+3)!$ permutations on the $j+3$ sites. Thus $\mathbf{xyz}$ of width

j appears with current width l with probability :

$$\frac{3 \binom{j}{l} (l+2)! \, (j-l)!}{(j+3)!} = \frac{3(l+1)(l+2)}{(j+1)(j+2)(j+3)}$$

The required probability is the sum of the last ones, for all l :

- if $j \geq k$, l can only be $< k$, so we get

$$\sum_{l=0}^{k-1} \frac{3(l+1)(l+2)}{(j+1)(j+2)(j+3)} = \frac{k(k+1)(k+2)}{(j+1)(j+2)(j+3)}$$

- if $j < k$

$$\sum_{l \leq j} \frac{3(l+1)(l+2)}{(j+1)(j+2)(j+3)} = \frac{(j+1)(j+2)(j+3)}{(j+1)(j+2)(j+3)}$$

$$= 1$$

which agrees with Property $(\mathcal{P})$: a triangle with width $< k$ will necessarily appear at some stage of the construction. □

Lemma 5.11 *Let* $x(yz)t$ *be a bicycle of width* j *such that* yzt *is a son or a stepson of* xyz. *The probability that such a bicycle appears during the construction is*

$$\frac{3!}{(j+1)(j+2)(j+3)(j+4)}$$

Proof. A bicycle $x(yz)t$ is always created with current width 0. Now, to get yzt as a son or a stepson of xyz, we need the following condition : t must be inserted after x, y and z ; thus there are 3! possibilities to insert x, y, z and t. Then the j sites inside the bicycle can be inserted in any of the $j!$ possible orders. So the probability is :

$$\frac{3! j!}{(j+4)!} = \frac{3!}{(j+1)(j+2)(j+3)(j+4)}$$

This result also agrees with property $(\mathcal{P})$: a bicycle with width 0 will always appear, and the last site inserted among x, y, z and t is t with probability $\frac{1}{4}$. □

The following lemma is a direct consequence of the bound on the size of higher order Voronoi diagrams. It gives in this case a bound for $|\mathcal{F}_j(\mathcal{S})|$, more precise that the general bound given by Theorem 2.3 for $|\mathcal{F}_{\leq j}(\mathcal{S})|$.

Lemma 5.12 *The number of triangles having width* j *is*

$$|\mathcal{F}_j(\mathcal{S})| \leq (2j+1)n$$

Proof. $|\mathcal{F}_j(\mathcal{S})|$ is exactly the number of close-type vertices of the order $j+1$ Voronoi diagram. This number is computed in [Lee82], which gives the result. $\square$

We propose here an alternative proof for Lemma 4.10, owing to the preceding lemma.

Lemma 5.13 *The number of bicycles having width at most j is*

$$|\mathcal{G}_{\leq j}(\mathcal{S})| = O(n(j+1)^3)$$

Proof. We first notice that a given segment joining two points of

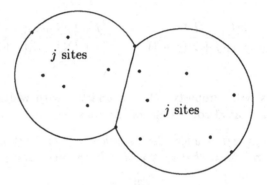

Figure 5.6 : For the proof of Lemma 5.13

$\mathcal{S}$ is an edge of at most $2j+2$ triangles of width less than j : in Figure 5.6, every triangle defined by E as an edge, and a site lying outside the two drawn circles will be of width superior to j, and the sites lying on or inside these circles allow us to form at most $2j+2$ triangles of width at most j.

We then bound the number of bicycles of width $\leq j$ by subdividing a bicycle into two adjacent triangles, one of width l and the other of width less than j.

$$
\begin{aligned}
|\mathcal{G}_{\leq j}(\mathcal{S})| &\leq \sum_{l=0}^{j} |\mathcal{F}_l(\mathcal{S})| 3(2j+2) \\
&\leq 6(j+1) \sum_{l=0}^{j} O(n(l+1)) \\
&\leq 6(j+1) O(n(j+1)^2) \\
&= O(n(j+1)^3)
\end{aligned}
$$

$\square$

5.2.2   Analysis of the expected space used by the k-Delaunay Tree

Lemma 5.14 *The expected number of nodes in the k-Delaunay Tree is $O(k^2 n)$.*

Proof. This number is the number of all successive vertices arising in all order $\leq k$ Voronoi diagrams during the construction. It is less than

$$\sum_{j=0}^{n-3} \sum_{S \in \mathcal{F}_j(S)} \mathrm{Prob}(S \text{ arises during the construction})$$

$$= \sum_{j=0}^{k-1} \sum_{S \in \mathcal{F}_j(S)} 1 + \sum_{j=k}^{n-3} \sum_{S \in \mathcal{F}_j(S)} \frac{k(k+1)(k+2)}{(j+1)(j+2)(j+3)}$$

$$\text{(using Lemma 5.10)}$$

$$= \sum_{j=0}^{k-1} |\mathcal{F}_j(S)| + k(k+1)(k+2) \sum_{j=k}^{n-3} \frac{|\mathcal{F}_j(S)|}{(j+1)(j+2)(j+3)}$$

$$= O(k^2 n) \text{ , using Lemma 5.12 and } \sum_{j=k}^{n-3} \frac{1}{(j+2)(j+3)} = O(\frac{1}{k})$$

$\square$

We already know that the total number of links to sons and stepsons is less than the number of nodes (Lemma 5.6), and that each node has at most 6 neighbors. We conclude :

Proposition 5.15 *The expected space complexity of the k-Delaunay Tree is $O(k^2 n)$.*

5.2.3   Analysis of the expected cost of Procedure location

We want here to count the number of nodes visited during the construction. During the insertion of **m**, Procedure location looks at first for a Delaunay triangle in conflict with **m**. To this end, a triangle **yzt** may be visited if **yzt** is the son or the stepson of a triangle **xyz** such that $m \in B(\mathbf{xyz})$. So a bicycle $\mathbf{x(yz)t}$ of width j obliges to visit up to j nodes. Secondly, Procedure propagate is called, and a triangle is visited if it is the neighbor of a triangle **xyz** such that $m \in B(\mathbf{xyz})$, and **xyz** is not dead. So a triangle of width j obliges to visit up to $6 \inf(j, k-1)$ nodes.

Lemma 5.16 *The expected total number of visited nodes during the construction is $O(n \log n + k^3 n)$.*

Proof. The total number of visited nodes during the search for a triangle of the Delaunay triangulation in conflict with a new site is inferior to :

$$\sum_{j=0}^{n-4} \sum_{\mathbf{x(yz)}t\in\mathcal{G}_j(\mathcal{S})} j\operatorname{Prob}\left(\begin{array}{c} \mathbf{yzt} \text{ arises as a son} \\ \text{or a stepson of } \mathbf{xyz} \\ \text{during the construction} \end{array}\right)$$

We then bound this expression by $O(n\log n)$, using Lemmas 5.11 and 5.13, in a way that is similar to the proof of Lemma 4.1. When such a triangle has been found, Procedure **propagate** then explores all triangles in conflict with the site, using the neighborhood relationships. The total number of triangles visited during the propagation is less than :

$$\sum_{j=0}^{n-3} \sum_{S\in\mathcal{F}_j(\mathcal{S})} 6\inf(j, k-1)\operatorname{Prob}\left(S \text{ arises during the construction}\right)$$

Similar calculations prove that this is $O(k^3 n)$, using Lemma 5.10. $\square$

Since it takes constant time to process a node, Lemma 5.16 yields the following proposition :

Proposition 5.17 *The expected cost of Procedure* location *is* $O(n\log n + k^3 n)$.

5.2.4 Analysis of the expected cost of Procedure creation

This cost is the cost of creating all the vertices of the order $\leq k$ Voronoi diagrams, plus the cost of maintaining the adjacency relationships, after each insertion of a new site m. If $x(\mathbf{m})$ is the number of new triangles created after the insertion of m, the first cost is $O(x(\mathbf{m}))$, since creating a triangle costs a constant time. As we have seen in Section 5.1.3.5, the second cost is $O(x(\mathbf{m})\log k)$. The overall cost is thus, using Lemma 5.14 :

$$O\left(\sum_{\mathbf{m}} x(\mathbf{m})\right) + O\left(\sum_{\mathbf{m}} x(\mathbf{m})\log k\right) \leq O(k^2 n\log k)$$

As mentioned in Remark 5.8, this could be improved to $O(k^2 n)$ complexity.

Proposition 5.18 *The expected cost of Procedure* creation *is* $O(k^2 n\log k)$.

Altogether, Propositions 5.15, 5.17 and 5.18 prove Theorem 5.9.

5.3 *l*-nearest neighbors

The k-Delaunay Tree contains the combinatorial structure of any order l Voronoi diagram ($l \leq k$). We show in the following section how such a diagram can be extracted from the k-Delaunay Tree.

As shown by the analysis of Procedure location, the k-Delaunay Tree is an efficient data structure to perform point location. Section 5.3.2 shows that it can be readily used to search the l nearest neighbors of a given point.

5.3.1   Deducing the order l Voronoi diagram from the k-Delaunay Tree ($l \leq k$)

Theorem 5.19 $Vor_l(S)$ ($l \leq k$) can be deduced from the k-Delaunay Tree in time proportional to the size of $Vor_l(S)$, which is $O(ln)$.

> *Proof.* Remember that the k Delaunay Tree maintains all the adjacency relations in $Vor_l(S)$. So we can find $Vor_l(S)$ in time proportional to its size, as soon as we know one of its vertices. We thus only have to find one vertex of each Voronoi diagram.
>
> Assume we know an infinite triangle $pq\infty$ of current width 0. Its finite edge $[pq]$ is necessarily an edge of the current convex hull. Let s_0 be the site such that pqs_0 is a neighbor of $pq\infty$ and that $B(pqs_0)$ is empty. Let $s_1, \ldots, s_{k-1}$ be the sites such that triangle pqs_i is the including neighbor of triangle pqs_{i-1}, $i = 1, \ldots, k - 1$. Such sites always exist (provided that $k < n - 1$) since $[pq]$ is an edge of the convex hull of the already inserted sites. Each triangle pqs_i is associated to a node in the k-Delaunay Tree and its current width is i. Triangle pqs_i is dual to a vertex ν_i of $Vor_{i+1}(S)$.
>
> If we maintain a pointer to an infinite triangle of current width 0 (which can be done in constant time at each stage of the construction), we can find the s_i, $i \leq l$, and thus a vertex of each $\leq l$ Voronoi diagrams, in time $O(l)$.   □

> *Remark 5.20* Now, suppose we want to label the regions of $Vor_i(S)$, $i \leq k$, by the i nearest sites of an (arbitrary) point of the interior of that region. We remark that the label of one region of $Vor_{i+1}(S)$ incident to ν_i is $\{s_0, s_1, \ldots, s_i\}$. As soon as we know the label of one particular region, we can deduce the labels of all the other regions by traversing the diagram. Each time we cross an edge E, we come out of one region, say C, and enter another region, say C'. The labels of C and C' differ by only one site. The site of the label of C which is not in the label of C', and the site of the label of C' which is not in the label of C, are precisely the two sites whose bisector supports E. We have to substitute this first site by the second one in the label of C to get the label of C'.

5.3.2   Finding the l nearest neighbors

Let us consider now the problem of finding the l nearest neighbors ($l \leq k$) of a given point. This problem is equivalent to finding the label of the region $V(\{p_1, p_2, \ldots, p_l\})$ of $Vor_l(S)$ which contains point m. Procedure location gives the vertices of the Voronoi diagrams whose associated disk contains $\mathbf{m}$. It remains to show how to find the label of a region. This point must be clarified since our structure represents vertices of the Voronoi regions and we know, from Lemma 5.1, that the label of a Voronoi polygon is not included in the union of the labels of its vertices and thus, cannot be deduced from them (the label of a vertex consists of the three sites which are the vertices of its dual triangle ; it is also the symmetric difference of the labels of its incident regions).

However, we can take advantage of the fact that we know all the order $\leq l$ Voronoi diagrams to compute the label of a region. The following lemma will help in that task.

Lemma 5.21 *Let $\mathbf{m} \in V(\{\mathbf{p}_1, \mathbf{p}_2, \ldots, \mathbf{p}_l\})$ in $Vor_l(S)$, where $\mathbf{p}_1, \mathbf{p}_2, \ldots, \mathbf{p}_l$ are some sites of S (with $\delta(\mathbf{p}_i, \mathbf{m}) \leq \delta(\mathbf{p}_j, \mathbf{m})$ if and only if $i \leq j$). Then, for each $i \in \{1, \ldots, l\}$, there exists a vertex ν_i of $Vor_i(S)$, such that ν_i is dual to a triangle S_i having $\mathbf{p}_i$ as a vertex, and such that $\mathbf{m} \in B(S_i)$.*

Proof. Since $\mathbf{p}_l$ is a lth nearest site from $\mathbf{m}$, we have $\delta(\mathbf{p}_i, \mathbf{m}) \leq \delta(\mathbf{p}_l, \mathbf{m})$, $\forall i = 1, \ldots, l$.

Let $\mathbf{q}$ be a point on the line $(\mathbf{m}\mathbf{p}_l)$: $\mathbf{q} = t\mathbf{m} + (1-t)\mathbf{p}_l, t \in \mathbb{R}$.

$$\delta(\mathbf{p}_l, \mathbf{q}) = |t|\delta(\mathbf{p}_l, \mathbf{m})$$
$$\delta(\mathbf{m}, \mathbf{q}) = |1-t|\delta(\mathbf{p}_l, \mathbf{m})$$

Let us suppose that $t \geq 1$, i.e. $\mathbf{m}$ lies between $\mathbf{q}$ and $\mathbf{p}_l$.

The following inequalities hold, for $i = 1, \ldots, l$:

$$\begin{aligned} \delta(\mathbf{p}_i, \mathbf{q}) &\leq & \delta(\mathbf{p}_i, \mathbf{m}) + \delta(\mathbf{m}, \mathbf{q}) \\ &\leq & \delta(\mathbf{p}_l, \mathbf{m}) + (t-1)\delta(\mathbf{p}_l, \mathbf{m}) \\ &= & t\delta(\mathbf{p}_l, \mathbf{m}) \\ &= & \delta(\mathbf{p}_l, \mathbf{q}) \end{aligned}$$

So, if we move $\mathbf{q}$ on the half line $D^+_{\mathbf{m}\mathbf{p}_l} = \{t\mathbf{m} + (1-t)\mathbf{p}_l, t \geq 1\}$ supported by $(\mathbf{m}\mathbf{p}_l)$, a furthest site from $\mathbf{q}$ among $\{\mathbf{p}_1, \mathbf{p}_2, \ldots, \mathbf{p}_l\}$ remains $\mathbf{p}_l$, the same as from $\mathbf{m}$. We can deduce that the intersection of $D^+_{\mathbf{m}\mathbf{p}_l}$ with the boundary of $V(\{\mathbf{p}_1, \mathbf{p}_2, \ldots, \mathbf{p}_l\})$ is a point $\mathbf{q}$ belonging to the bisecting line $Bis(\mathbf{p}_l, \mathbf{p}')$ of $[\mathbf{p}_l, \mathbf{p}']$, where $\mathbf{p}'$ is a site of S not in $\{\mathbf{p}_1, \mathbf{p}_2, \ldots, \mathbf{p}_l\}$. Let us denote by F the edge of $V(\{\mathbf{p}_1, \mathbf{p}_2, \ldots, \mathbf{p}_l\})$ supported by $Bis(\mathbf{p}_l, \mathbf{p}')$ (see Figure 5.7).

Let us define $R = \{r \in Bis(\mathbf{p}_l, \mathbf{p}')/\mathbf{m} \in C(r, \mathbf{p}_l)\}$ where $C(r, \mathbf{p}_l)$ denotes the disk of center r and radius $\delta(r, \mathbf{p}_l)$. R is a half line of $Bis(\mathbf{p}_l, \mathbf{p}')$ containing $\mathbf{q}$. So $F \cap R \neq \emptyset$, which implies that one of the two end points of F, say ν_l, belongs to R. ν_l is a vertex of $Vor_l(S)$ dual to a triangle S_l having $\mathbf{p}_l$ as one of its vertices, and whose disk contains $\mathbf{m}$.

The lemma is thus proved for $\mathbf{p}_l$. For the other $\mathbf{p}_i, i = 1, 2, \ldots, l-1$, the same proof still holds, considering $Vor_i(S)$ instead of $Vor_l S$). $\square$

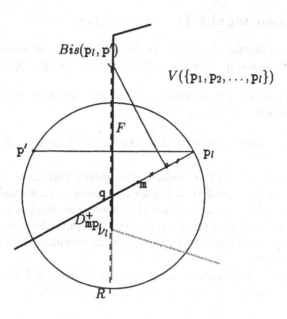

Figure 5.7 : For the proof of Lemma 5.21

If we want to find the label of the region of $Vor_l(\mathcal{S}), l \leq k$, which contains m, we can locate it, by applying Procedure location to the l-Delaunay Tree. We first find all the triangles whose balls contain m. We then have to find, among the vertices of those triangles, the l nearest sites from m, which can be done in a straightforward way.

5.4   The k-Delaunay Tree in higher dimensions

In this section, we generalize the previous results to higher dimensions. Let $\mathcal{S}$ be a set of n sites in d-space such that no subset of $d+2$ sites lie on a same sphere and no subset of $d+1$ sites are coplanar.

The notion of including and excluding neighbors can be generalized. A simplex has $d+1$ excluding neighbors and $d+1$ including ones, one through each of its facets. To any pair of adjacent simplices correspond an edge in some higher order Voronoi diagram. If ν is a close-type vertex (resp. far-type vertex) of $Vor_k(\mathcal{S})$, the edges of $Vor_k(\mathcal{S})$ issued from ν correspond to the neighborhood relationships between the simplex dual to ν and its excluding (resp. including) neighbors. If ν is a medium-type vertex (see section 1.2.2), the edges of $Vor_k(\mathcal{S})$ issued from ν correspond to both types of neighborhood relationships.

5.4.1 The d dimensional k-Delaunay Tree

We can generalize the structure developed in Section 5.1.3. The d-dimensional Delaunay Tree is a direct acyclic graph satisfying the following property :

> ($\mathcal{P}$) *all the simplices of current width strictly less than k are present in the d-dimensional k-Delaunay Tree.*

The extension of the construction of the k-Delaunay Tree to higher dimensions is straightforward.

The technique of Section 5.3.1 that deduces the order l Voronoi diagram from the k-Delaunay Tree can be extended to higher dimensions in a straightforward manner. As in Section 5.3.1, we can traverse the graph consisting of the vertices and the edges of the order l Voronoi diagram and compute the labels of each region. From this graph and the labels, it is plain to compute the faces of all dimensions of the diagram.

The algorithm presented in Section 5.3.2 can also be extended without difficulty, to compute the l nearest neighbors of a given site ($l \leq k$).

5.4.2 Analysis of the randomized construction

This section presents the following theorem, which generalizes Theorem 5.9 to d dimensions :

Theorem 5.22 *For any set of n sites, if we randomize the sequence of their insertions, the k-Delaunay Tree (and thus the order $\leq k$ Voronoi diagrams) of the n sites can be constructed in expected time $O\left(k^{\lceil \frac{d+1}{2} \rceil + 1} n^{\lfloor \frac{d+1}{2} \rfloor}\right)$ using expected storage $O\left(k^{\lceil \frac{d+1}{2} \rceil} n^{\lfloor \frac{d+1}{2} \rfloor}\right)$.*

Our bounds are not as good as those of K. Mulmuley [Mul91a] : the increase by one in the exponent of k is a consequence of the fact that we maintain, at each stage of the incremental insertion, the complete informations relative to all order $\leq k$ Voronoi diagrams, whereas Mulmuley only maintains the vertices of the diagrams, without maintaining their order (which can be deduced at the end of the construction).

We will see in Chapter 7 the works of K. Mulmuley concerning dynamic algorithms.

We only give the main steps achieving the proof of Theorem 5.22.

As in Section 5.2.3, we can define, for $d+2$ sites $x_1, x_2, \ldots, x_{d+2}$, the bicycle $x_1(x_2 \ldots x_{d+1})x_{d+2}$. We also define the width of a simplex and the width of a bicycle.

The following lemmas generalize Lemmas 5.10 and 5.11.

Lemma 5.23 *Let $x_1 x_2 \ldots x_{d+1}$ be a simplex having width j. $x_1 x_2 \ldots x_{d+1}$ will arise as a vertex of some order $\leq k$ Voronoi diagram during the construction* *with probability :*

$$\begin{cases} \dfrac{k(k+1)\ldots(k+d)}{(j+1)(j+2)\ldots(j+d+1)} & \text{if } j \geq k \\ 1 & \text{if } j < k \end{cases}$$

Lemma 5.24 *Let* $x_1(x_2\ldots x_{d+1})x_{d+2}$ *be a bicycle of width* j *such that* $x_2x_3\ldots$ x_{d+2} *is a son or a stepson of* $x_1x_2\ldots x_{d+1}$. *The probability that such a bicycle appears during the construction is*

$$\frac{(d+1)!}{(j+1)\ldots(j+d+2)}$$

Applying Theorem 2.3, we have :

Lemma 5.25 *The number of simplices having width at most* j *is*

$$|\mathcal{F}_{\leq j}(\mathcal{S})| = O\left(n^{\lfloor \frac{d+1}{2} \rfloor}(j+1)^{\lceil \frac{d+1}{2} \rceil}\right)$$

Lemma 5.26 *The number of bicycles having width at most* j *is*

$$|\mathcal{G}_{\leq j}(\mathcal{S})| = O\left(n^{\lfloor \frac{d+1}{2} \rfloor}(j+1)^{\lceil \frac{d+1}{2} \rceil + 1}\right)$$

Let us now compute the complexity of the k-Delaunay Tree.

Proposition 5.27 *The expected space complexity of the k-Delaunay Tree is*

$$O\left(k^{\lceil \frac{d+1}{2} \rceil}n^{\lfloor \frac{d+1}{2} \rfloor}\right)$$

Proof. The proof is similar to the one of Proposition 5.15, using Lemmas 5.23 and 5.25, and the fact that, as in 2 dimensions, the total number of links to sons and stepsons is less than the number of nodes (Lemma 5.6), and each node has at most $2(d+1)$ neighbors. $\square$

Proposition 5.28 *The expected cost of Procedure* location *is*

$$O\left(k^{\lceil \frac{d+1}{2} \rceil + 1}n^{\lfloor \frac{d+1}{2} \rfloor}\right)$$

Proof. Lemmas 5.24 and 5.26 allow to prove the result, using arguments similar to those used in the proof of Lemma 5.16. $\square$

The cost of Procedure creation is the cost of creating all the vertices of the order $\leq k$ Voronoi diagrams plus the cost of maintaining the neighborhood relationships, after each insertion of a new site m. The computation of the neighborhood relationships between the new simplices created by the insertion of a new point m is the same as in the two dimensional case, an edge is simply replaced by a $(d-1)$-face.

So we obtain :

Proposition 5.29 *The expected cost of Procedure* creation *is*

$$O\left(k^{\lceil\frac{d+1}{2}\rceil}n^{\lfloor\frac{d+1}{2}\rfloor}\log k\right)$$

Altogether, Propositions 5.27, 5.28 and 5.29 prove Theorem 5.22.

5.5 Experimental results

It is to be noted that the algorithm is simple, even if its description and its analysis may look rather intricate ! The core of the algorithm is given in Figures 5.3 and 5.4. Moreover, the numerical computations involved are also quite simple : they consist mostly of comparisons of (squared) distances in order to check if a point lies inside or outside a ball. The algorithm has been implemented in the two dimensional case. The program consists of less than 1000 lines of C. It has run on many examples with different kinds of point distributions. Some results and statistics are presented in Figures 5.8 to 5.16.

5.5.1 Influence of randomization

In Figure 5.8, the points lie on three ellipses, two for the *eyes* and one for the *head*. We tried several permutations of the points for the computation of the 3-Delaunay Tree :

(1) Points of the head in the order along the ellipse, and then points in the order along each eye.

(2) The same as the preceding, except that one point of an eye is inserted first.

(3) Points on the eyes in order, and then points on the head in order.

(4) A random permutation.

(5) Another random permutation.

In Figure 5.9, the function drawn in dotted line shows the total number of vertices of the order ≤ 3 Voronoi diagrams, versus the number of inserted sites. The functions drawn in bold line show the numbers of nodes in the 3-Delaunay Tree, for the different permutations. The functions drawn in thin line show the numbers of nodes visited by the first part of Procedure location (we will call

Figure 5.8 : Delaunay triangulation and order 3 Voronoi diagram of a set of 415 points

those nodes *1-visited nodes* for short, since they correspond to a traversal of the *(1-)Delaunay Tree*), to find a Delaunay triangle in conflict with the new point.

Figure 5.10 gives some statistics about the computation using the different permutations : the size of the 3-Delaunay Tree and the sum of the sizes of the order ≤ 3 Voronoi diagrams at the end of the execution, and for the insertion of one point, the maximal and average numbers of 1-visited nodes and created nodes.

This example shows that, if degenerate orders may be really unefficient, the behaviour for random order, or even for the third permutation is satisfying.

5.5.2   Influence of k

To study the effect of increasing k on the algorithm, we use the set of points depicted in Figure 5.11.

Figure 5.12 presents, on the left side, the size of the (1-)Delaunay Tree in bold line, the size of the (order 1) Voronoi diagram in dotted line and in thin line the number of nodes that are 1-visited by Procedure location, which does not depend on k. The right side of Figure 5.12 shows the size of the k-Delaunay Tree and the sum of the sizes of the order $\leq k$ Voronoi diagrams for $k = 1,2,3,4,6$ and 10.

Figure 5.13 presents statistics similar to those in Figure 5.10.

5.5.3   Influence of the point distribution

These last statistics concern the influence of the point distribution. To this aim, we compute the 3-Delaunay Tree for the four sets of 400 points described in Figures 5.8, 5.11 and 5.14. Figure 5.15 shows in this left part the size of the 3-Delaunay Tree, and the sum of the sizes of the order ≤ 3 Voronoi diagrams and

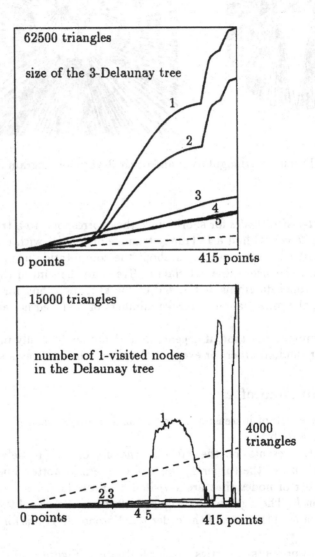

Figure 5.9 : Results for the set of points of Figure 5.8, $k = 3$, different permutations for the insertion

Permutation	1	2	3	4	5
Size of the 3-Delaunay Tree	62522	48334	14958	10903	10395
Size of the ≤ 3 Voronoi	3917	3917	3917	3917	3917
Max nb of 1-visited nodes	14835	1245	1329	181	73
Average nb of 1-visited nodes	2123	245	239	33.5	27
Max nb of created triangles	1044	853	71	210	179
Average nb of created triangles	151	117	36	26.4	25.1

Figure 5.10 : Statistics for the set of points of Figure 5.8

the number of 1-visited nodes for the four sets. This figure demonstrates that, with a randomization of the input data, the average behaviour of the algorithm does not depend on the distribution of the points. The better result for points of Figure 5.8 is only due to the smallest size of the output (which is related to the points on the k-hulls). One can see also that the location time is very small in comparison with the sum of the sizes of the ≤ 3 Voronoi diagrams. The right part of Figure 5.15 shows only the location time for the forty last points. Figure 5.16 presents statistics as in Figures 5.10 and 5.13.

The experimental results show that the algorithm performs well even on degenerate distributions of points.

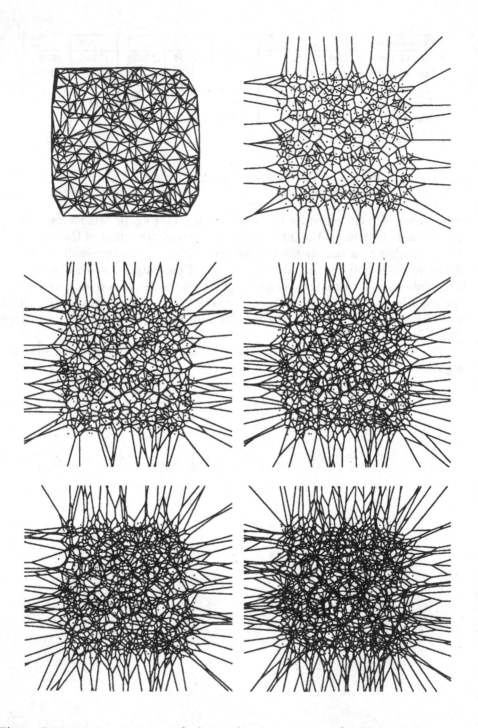

Figure 5.11 : Delaunay triangulation and order 1,2,3,4 and 6 Voronoi diagrams of a set of 400 random points in a square

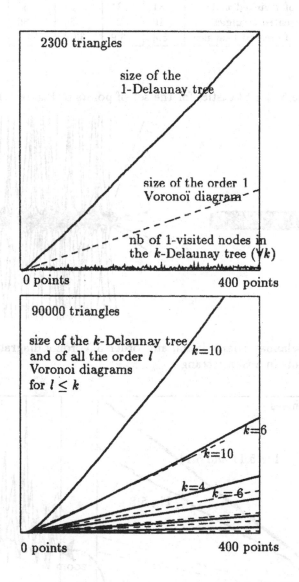

Figure 5.12 : Results for the set of points of Figure 5.11, for different values of k

k	1	2	3	4	6
Size of the *k*-Delaunay Tree	2307	6748	13246	21694	43740
Size of the ≤ *k* Voronoi	799	2378	4720	7815	16198
Max nb of 1-visited nodes	79	79	79	79	79
Average nb of 1-visited nodes	31	31	31	31	31
Max nb of created triangles	10	29	50	80	150
Average nb of created triangles	5.8	16.9	33.2	54.5	110

Figure 5.13 : Statistics for the set of points of Figure 5.11

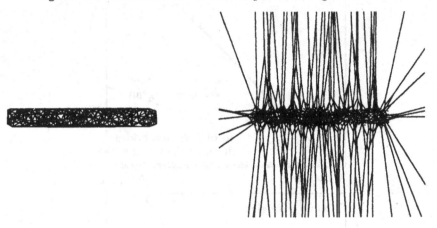

Figure 5.14 : Delaunay triangulation and order 3 Voronoi diagram of a set of 400 random points in a thin rectangle

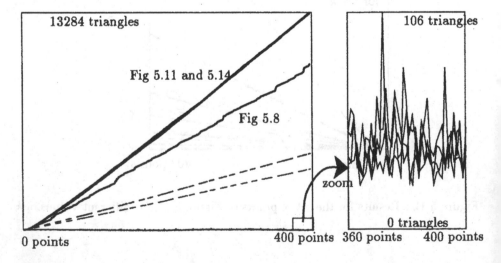

Figure 5.15 : Results for the set of points of Figures 5.8, 5.11 and 5.14 for $k = 3$

Set of points	Fig. 7	Fig. 10	Fig. 13	Fig. 14
Size of the 3-Delaunay Tree	10141	13246	13284	13184
Size of the ≤ 3 Voronoi	3776	4720	4718	4718
Max nb of 1-visited nodes	109	79	90	78
Average nb of 1-visited nodes	30	31	33	34
Max nb of created triangles	236	50	67	57
Average nb of created triangles	25.6	33.2	33.1	33.1

Figure 5.16 : Statistics for the sets of points of Figures 5.8, 5.11 and 5.14

Conclusion

We have shown that the k-Delaunay Tree of n points can be constructed in $O(n \log n + k^3 n)$ (resp. $O\left(k^{\lceil \frac{d+1}{2} \rceil + 1} n^{\lfloor \frac{d+1}{2} \rfloor}\right)$) randomized expected time in the plane (resp. in d space). Its randomized expected size is $O(k^2 n)$ (resp. $O\left(k^{\lceil \frac{d+1}{2} \rceil} n^{\lfloor \frac{d+1}{2} \rfloor}\right)$). The k-Delaunay Tree allows to compute the order $\leq k$ Voronoi diagrams of n points within the same bounds. Any order $l \leq k$ Voronoi diagram can be deduced from the k-Delaunay Tree in time proportional to its size, which is $O(ln)$ in two dimensions. Moreover, the k-Delaunay Tree can be used to find the l nearest neighbors of a given point.

The structure could be extended to deal with k-levels of general arrangements. But in this case, it is not sufficient to store the Delaunay Tree augmented with additional neighborhood relations, since the hyperplanes are not supposed to have one face on the lowest level. We must store parent pointers between old and new triangles of all widths.

An important point is that these results hold whatever the point distribution may be.

The algorithm is simple and, moreover, the numerical computations involved are also quite simple : they consist mostly of comparisons of (squared) distances in order to check if a point lies inside or outside a ball. Experimental results, for uniform as well as degenerate distributions of points, have provided strong evidence that this algorithm is very effective in practice, for small values of k.

For large values of k, a similar structure, based on the order k furthest neighbors Voronoi diagrams, could be derived. It would provide results similar to the ones above to construct all order $\geq n - k$ Voronoi diagrams and to find l furthest neighbors for $l \leq k$.

Chapter 6

Towards a fully dynamic structure

We describe here the dynamization of the Influence Graph, on two examples :
the Delaunay triangulation of point sites in any dimension, and the arrangement
of line segments in the plane. In both cases, the main idea is the same : when
an object is removed from the structure, we decide to "reconstruct the past", as
if this object had never existed.

However, in spite of this similarity, the details of the algorithms differ a lot,
as will be seen, and it seems to be difficult to design a general —and precise,
too— framework for dealing with deletion of an object.

[CMS92] uses the same idea, of restoring the history as if the object had never
been inserted, for convex hulls in any dimension. [Sch91] computes the whole
history of insertions, but also deletions, and in this way gets a bigger history
that he has to clean up sometimes. [Mul91b] uses a different technique which
does not involve the history of the construction. All these approaches will be
rapidly presented in Chapter 7.

6.1   Removing a site from the Delaunay triangulation

In this section, we describe an algorithm maintaining the Delaunay Tree under
insertions and deletions of sites. This can be done in $O(\log n)$ expected time for
an insertion and $O(\log \log n)$ expected time for a deletion in the plane [DMT92a],
where n is the number of sites currently present in the structure. For deletions,
by expected time, we mean averaging over all already inserted sites for the choice
of the deleted sites. The algorithm has been effectively coded and experimental
results are given. The method extends to higher dimensions.

Let S be a set of n sites in the euclidean plane, such that no four sites are
cocircular.

We assume that the Delaunay Tree has been constructed for the set S, by
using the incremental randomized algorithm (see Chapter 3). We now want to
remove a site p of S. All the triangles incident to p must be removed from the
Delaunay Tree : some of them are triangles of the Delaunay triangulation of S
(so they are leaves of the Delaunay Tree), but other ones already died ; they
correspond to internal nodes of the Delaunay Tree, and must be removed, too.
Moreover, we must restore the Delaunay Tree in the same state it would be in if
p had never been inserted, and if the other sites had been inserted in the same
order. That way, we preserve the randomized hypothesis on the sequence of
sites, and the conditions for further insertions or deletions are fulfilled.

We must thus reconstruct a past for the final triangulation in which p takes
no part. The deletion of p creates a "hole" in each successive triangulation after
the insertion of p, which the tree keeps a trace of. The idea of our algorithm is
to fill each hole with the right Delaunay triangulation.

Let us describe the structure of a node of the Delaunay Tree, obtained from
the structure given in Chapter 3 by adding a few new fields (some of them have
not been used yet and will be defined in the following) :

- the triangle : creator vertex, two other vertices, circumscribed circle
- a mark **dead**
- pointers to the at most three sons and the list of stepsons
- pointers to the father and the stepfather
- the three current neighbors if the triangle is not dead, the three neighbors at the death otherwise
- the three neighbors at the time of the creation
- two special neighbors
- a pointer **killer** to the site that killed the triangle
- a mark **to be removed**
- three pointers **star** to elements of structure **Star**

The insertion phase described in Chapter 3 can obviously maintain those additional informations.

6.1.1 Different kinds of modified nodes

Let us describe how the deletion of p affects the nodes in the Delaunay Tree.

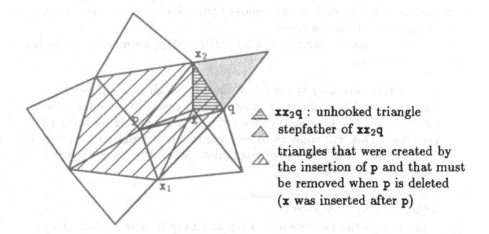

xx_2q : unhooked triangle

stepfather of xx_2q

triangles that were created by the insertion of p and that must be removed when p is deleted (x was inserted after p)

Figure 6.1 : The two kinds of modified nodes

Some nodes must be removed : they correspond to triangles having p as a vertex. Depending on its time of creation there are two cases for such a node : either it has been created by the insertion of p, or it has been created by the insertion of some site afterwards ; the latter occurs *iff* its father and stepfather both have p as a vertex and thus both the parents must be removed, too. During the construction of the Delaunay Tree, some sites did not create any triangle to be removed now, but if a site x created such a triangle, it created in fact two triangles to be removed : the two triangles created by x sharing edge px, see Figure 6.1. One of them, say xx_1p, is oriented clockwise, and the second one, xx_2p, is oriented counterclockwise.

A node to be deleted xx_2p may have a son (or a stepson) xx_2q that does not have p as a vertex and thus must remain in the Delaunay tree, see Figure 6.1. Such a node loses just one of its two parents and is therefore called *unhooked*. We must find a new parent in replacement of the lacking one.

The sketch of the method is the following :
Search step : Find all nodes of the Delaunay Tree that have to be removed, and all unhooked nodes
Reinsertion step : Locally reinsert the sites that are creators of the triangles found during the Search step, and update the triangulation

6.1.2 The Search step

By the discussion above, the set of nodes to be removed can be found by searching the Delaunay Tree starting from the nodes that were created by p. At each node marked to be removed we visit all its sons and stepsons recursively. If one of them has p as a vertex, it will be marked to be removed as well. Otherwise it is an unhooked node. The creator of both these types of triangles must be reinserted, in order to replace the removed triangles by other triangles, and to hang up unhooked triangles again.

In order to be able to perform the Reinsertion step, we must store the list of sites to be reinserted :

We need an auxiliary structure, **Reinsert**, which is a balanced binary tree consisting of the set of sites which created the nodes to be removed and the unhooked nodes ; the sites are sorted by order of insertion. This will allow us to reconstruct the triangles which will fill the holes in the successive triangulations, and to hang up again the unhooked nodes.

An element of **Reinsert** contains :
• the site x to be reinserted
• pointers to the two triangles xx_1p and xx_2p that were created by the insertion of x, if they exist (see Section 6.1.1). xx_1p is oriented clockwise
• the list of unhooked triangles that were created by the insertion of x

The search is initialized by the set C of nodes created by p.

To this aim, we must maintain an auxiliary array, **Created**, containing, for each site s of S, a pointer to one of the nodes created by s.

From this node, we can then compute the set C using the neighborhood relations at the time of creation and examining the creator of the triangles (Figure 6.2).

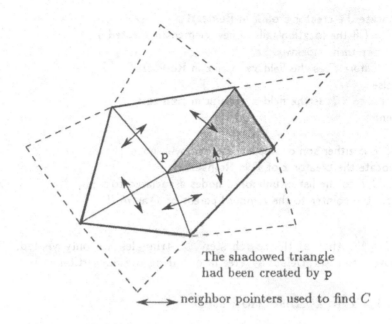

The shadowed triangle
had been created by p

⟷ neighbor pointers used to find C

Figure 6.2 : The search step : Initialization

```
for each element S of C
    for each son or stepson U of S
        examine(U)
    endfor ;
    remove links between S and its father and stepfather ;
    put S in "garbage collector"
endfor
```

We then simply recursively traverse the subgraph consisting of removed and unhooked nodes. Each son or stepson of a removed node is removed if it has **p** as a vertex and unhooked otherwise. All these nodes are added in the element of **Reinsert** associated to their creator. This process is detailed above.

```
examine(T)
if T is marked to be removed
    { T has already been visited }
    nothing
else
if p is a vertex of T
{ T = xsp}
    mark T to be removed ;
    for each son or stepson S of T
        examine(S)
    endfor ;
```

> locate the creator x of T in **Reinsert** ;
>> { if the location fails, a new element is created }
>
> if xsp turns clockwise
>> store T as the field xx_1p of x in **Reinsert**
>
> else
>> store T as the field xx_2p of x in **Reinsert**
>
> endif

else
> { T is either son or stepson of a removed node }
>
> locate the creator x of T in **Reinsert** ;
>
> add T to the list of unhooked nodes associated to x ;
>
> set the pointer to the removed parent of T as *null*

endif

Observe that in the search step the triangles are only visited. They are removed from the Delaunay Tree later during the reinsertion step.

6.1.3   The Reinsertion step

The sites contained in **Reinsert** must be reinserted in the Delaunay Tree in order to construct the successive triangulations without site **p**. The scheme of the reinsertion of a site **x** is the same as the usual scheme of insertion, except that everything happens locally : the location of a site **x** to be reinserted in the whole Delaunay Tree is unnecessary and would be too expensive.

The location in a generally small set A (for active) of triangles is sufficient. At the beginning of the reinsertion set A is initialized with all triangles killed by the insertion of **p**. They can be found by looking at the fathers of the triangles in C and following their neighbor pointers at their death.

Then, during the Reinsertion step, A is maintained so that it is the set of triangles in conflict with **p** in the Delaunay triangulation at the time just preceding the reinsertion of **x**. In each step, A is modified as follows : all the triangles of A in conflict with **x** are killed by **x** and thus disappear from the Delaunay triangulation and from A. The triangles created by the reinsertion of **x** appear in A (Figure 6.3), because they are in conflict with **p** (otherwise they would have existed in the triangulation containing **p**). The triangles of A not in conflict with **x** still remain in A. The triangles outside A are not modified by a reinsertion since they are not in conflict with **p** ; only their neighborhood or stepson relations involving removed nodes must be updated.

More precisely, the set A of triangles must be organized so that the location of conflicts is efficient. We can notice that the triangles in A form a star-shaped polygon with respect to **p**, since they are in conflict with **p**.

The edges and vertices (sites) of this polygon are stored in counterclockwise order in a circular list called **Star**. Note that the vertices of **Star** are the sites that are adjacent to **p**. Furthermore we maintain some pointers :

- Each edge of **Star** points to the adjacent triangle in A.

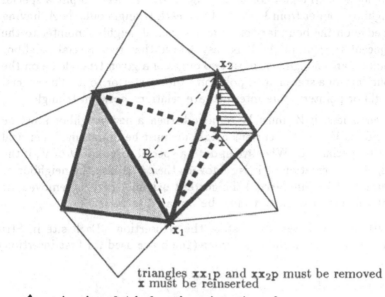

triangles xx_1p and xx_2p must be removed
x must be reinserted

triangles of A before the reinsertion of x
triangles remaining in A
new triangles of A after the reinsertion of x
an unhooked triangle

Figure 6.3 : A reinsertion

- Each triangle in A points to the adjacent edges of **Star** (at most three, pointers **star** in the description of a node).

- Each site which is a vertex of **Star** points to the edge of **Star** following the site.

Some elements of A are not represented in **Star**, but the whole set A can nevertheless be traversed using **Star** and pointers to neighbors.

Star can be initialized by first choosing a vertex of a triangle in A. We then follow the boundary of the star-shaped polygon using the neighborhood relations, and the pointers **star**.

We know the current neighbors of each triangle of A. Each edge e of **Star** is an edge of a triangle U of A and of a triangle V that does not belong to A.

Special neighbors :
The current neighbor of U through e is V, but the reciprocal relation does not always exist ; the neighbor pointer of V through e may reach another triangle W created a long time later.

If W must not be removed, this pointer must remain after the deletion of p. So we do not want to systematically modify the current neighbors of triangles not belonging to A. We need to put a special neighbor pointer from V to U. Thus each triangle outside A, having an edge on the boundary of A, has a special neighbor pointer to the adjacent triangle in A. It is easy to see that two special neighbor pointers are enough : at most two edges of a given triangle lie on the boundary of a star-shaped polygon, if it is exterior to it. The special neighbor pointers store intermediate relations between triangles.

Nevertheless, if W must be removed, then a new neighbor must be found for V, and the current neighbor must be maintained identical to the special one. When we update a special neighbor U of V, if the neighbor at creation of V is removed, then U is also the neighbor at creation of V. Similarly, if the current neighbor of V is removed, or if it belongs to A, then it must be updated to be U.

Everything is now set up to start the reinsertion. Each site in Structure **Reinsert** is reinserted in the right order (the order used for first insertion).

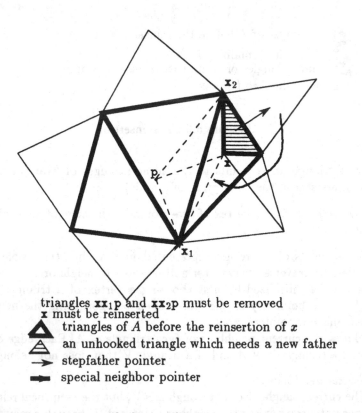

triangles $\mathbf{xx_1p}$ and $\mathbf{xx_2p}$ must be removed
$\mathbf{x}$ must be reinserted
triangles of A before the reinsertion of x
an unhooked triangle which needs a new father
stepfather pointer
special neighbor pointer

Figure 6.4 : An unhooked triangle with some removed triangles

Processing the unhooked triangles

Each element of **Reinsert** contains a site $\mathbf{x}$ to be reinserted, and the list of

corresponding unhooked triangles. To hang up such a triangle T again, we only
have to go to the remaining parent of it, which must have an edge in **Star**, and
then hang T up to the appropriate special neighbor of this parent. There may
also exist some removed triangles created by **x** (Figure 6.4). Notice that this
is not always true (Figure 6.5). If there is no removed triangle, the unhooked
triangle necessarily needs a stepfather, which is also the neighbor at creation
and the special neighbor, all these three triangles are set to the special neighbor
of the father.

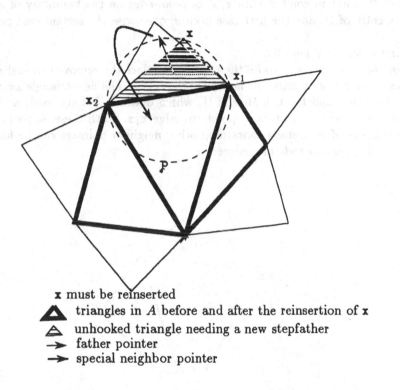

x must be reinserted

▲ triangles in A before and after the reinsertion of **x**

△ unhooked triangle needing a new stepfather

→ father pointer

➤ special neighbor pointer

Figure 6.5 : An unhooked triangle in the case that there is no removed triangle

Replacing the triangles to be removed by new ones

For each element **x** of **Reinsert**, we check if triangle **xx₁p** (and **xx₂p**) exists. If
xx₁p and **xx₂p** do not exist in the triangulation, then nothing has to be done.
Otherwise we have to fill the gap of triangles incident at **x** between edges **xx₁**
and **xx₂**. We must look at **Star** in order to find the triangles that have to be
created by the reinsertion of **x**. There are two cases : there may exist no triangle
of A in conflict with **x**, or several such triangles.

First note that **x₁** and **x₂** both belong to **Star** : in fact, before the insertion
of **x**, in the triangulation containing **p**, the edges **px₁** and **px₂** already existed
(by definition, the insertion of **x** created **xx₁p** and **xx₂p**) ; we know that the

vertices adjacent to p lie on **Star**. Let U be the triangle of A adjacent to the edge following x_1 on **Star** (remember that **Star** is oriented counterclockwise and xx_1p clockwise). U serves to distinguish between the two cases above. We have a direct access from x_1 to U, via edge xx_2.

After the reinsertion of x, the edges xx_1 and xx_2 will be on the boundary of the new set A of triangles in conflict with p. So, if there are some vertices on the current boundary of A, between x_1 and x_2, the triangles adjacent to the edges of this chain of vertices must be in conflict with x. U is such a particular triangle. Thus, if U is not in conflict with x, x_1x_2 is an edge on the boundary of A, and consequently of U, and the first case occurs, otherwise the second case occurs.

First case, see Figure 6.6 :
In this case, the only way to fill the gap is to replace the removed triangles xx_1p and xx_2p around x by only one new triangle xx_1x_2. The new triangle xx_1x_2 has U as stepfather and the neighbor of U, which does not belong to A, as father. x points to edge xx_2 of **Star**, x_1 points to edge x_1x. Both these edges point to triangle xx_1x_2 of A. Details concerning other neighbor pointers can be found in the following pseudo code procedure :

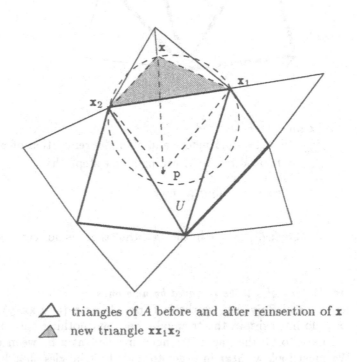

△ triangles of A before and after reinsertion of x
▲ new triangle xx_1x_2

Figure 6.6 : Reinsertion - removed triangles - First case

Reinsertion - removed triangles - First case
    create triangle xx_1x_2,

with U as stepfather and the neighbor of U through x_1x_2 as father ;
update the neighbor through x_1x_2 of U to be xx_1x_2 ;
find the neighbors at the creation of xx_1x_2, by looking at those of xx_1p and xx_2p ;
update the special neighbor pointers of the neighbors of xx_1x_2
 and their neighbor at creation, and current neighbor, if necessary ;
throw xx_1p and xx_2p away in "garbage collector" ;
add edges xx_1 and xx_2 in **Star** ;
add pointers from x to edge xx_2, and from x_1 to edge x_1x ;
update **Star** by letting xx_1 and xx_2 be incident to xx_1x_2 ;
put two **star** pointers from xx_1x_2 to the elements xx_1 and xx_2 of **Star** ;
in **Created**, xx_1x_2 is a triangle created by x

Second case, see Figures 6.7 and 6.8 :
We know that U is in conflict with x. We must find all the triangles in A in
conflict with x. Those triangles may be fathers for the nodes that will be created
by x. They form a connected subset of A, so they will be found owing to neighbor
pointers in the following way :

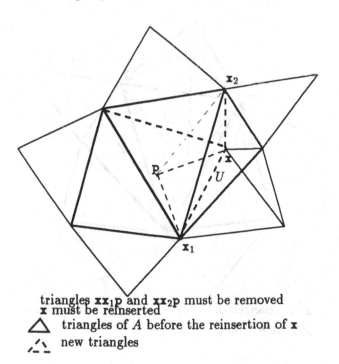

triangles xx_1p and xx_2p must be removed
x must be reinserted
△   triangles of A before the reinsertion of x
△   new triangles

Figure 6.7 : Reinsertion - removed triangles - Second case

Starting with U we visit the triangles in A incident at x_1 in counterclockwise
order until we reach a triangle not in conflict with x. Let V denote the last

vertices adjacent to p lie on **Star**. Let U be the triangle of A adjacent to the edge following x_1 on **Star** (remember that **Star** is oriented counterclockwise and xx_1p clockwise). U serves to distinguish between the two cases above. We have a direct access from x_1 to U, via edge xx_2.

After the reinsertion of x, the edges xx_1 and xx_2 will be on the boundary of the new set A of triangles in conflict with p. So, if there are some vertices on the current boundary of A, between x_1 and x_2, the triangles adjacent to the edges of this chain of vertices must be in conflict with x. U is such a particular triangle. Thus, if U is not in conflict with x, x_1x_2 is an edge on the boundary of A, and consequently of U, and the first case occurs, otherwise the second case occurs.

First case, see Figure 6.6 :
In this case, the only way to fill the gap is to replace the removed triangles xx_1p and xx_2p around x by only one new triangle xx_1x_2. The new triangle xx_1x_2 has U as stepfather and the neighbor of U, which does not belong to A, as father. x points to edge xx_2 of **Star**, x_1 points to edge x_1x. Both these edges point to triangle xx_1x_2 of A. Details concerning other neighbor pointers can be found in the following pseudo code procedure :

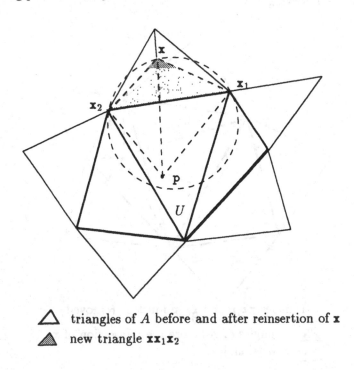

△ triangles of A before and after reinsertion of x

▲ new triangle xx_1x_2

Figure 6.6 : Reinsertion - removed triangles - First case

Reinsertion - removed triangles - First case
    create triangle xx_1x_2,

while the neighbor of V sharing vertex **s** is in conflict with **x**

    { we turn around **s** counterclockwise }

    $V \leftarrow$ this neighbor ;

    the **killer** of V is **x**

endwhile ;

$e \leftarrow$ edge of V through which we stopped finding conflict ;

create W' with edge e and vertex **x** ;

W' is the son of V, and the stepson of the neighbor of V through e ;

the neighbor at creation of W' through e is its stepfather ;

if e is an edge of **Star**

    update the element of e in **Star** by replacing V by W' ;

    let the **star** pointer of W' reach e ;

    put a pointer from **s** to e ;

    update the special neighbor of the neighbor of V through e to be W'

        and the ordinary neighbor if necessary

else

    update the neighbor of the neighbor of V through e to be W'

endif ;

if **s** $\neq$ **x**$_1$

    W' and W are neighbors and neighbors at creation through edge **xs**

endif ;

if **s** $=$ **x**$_1$

    $W_1 \leftarrow W'$;

    $W \leftarrow$ neighbor at creation of **xx**$_1$**p** through **xx**$_1$;

    W' and W are neighbors at creation through edge **xx**$_1$;

    the ordinary neighbor through **xx**$_1$ of W_1 is W ;

    the special neighbor through **xx**$_1$ of W is W_1 ;

    the ordinary neighbor through **xx**$_1$ of W is W_1 if necessary

endif ;

s $\leftarrow$ the other vertex of e ;

if **s** $=$ **x**$_2$

    $W_2 \leftarrow W'$;

    $W \leftarrow$ neighbor at creation of **xx**$_2$**p** through **xx**$_2$;

    W' and W are neighbors at creation through edge **xx**$_2$;

    the ordinary neighbor through **xx**$_2$ of W_2 is W ;

    the special neighbor through **xx**$_2$ of W is W_2 ;

    the ordinary neighbor through **xx**$_2$ of W is W_2 if necessary

endif ;

$W \leftarrow W'$

    { W must be stored for future neighborhood relations }

until **s** $=$ **x**$_2$;

store for example W_1 as created by **x** in **Created** ;

throw **xx**$_1$**p** and **xx**$_2$**p** away in "garbage collector" ;

repeat

    while the neighbor of V sharing vertex **s** is in conflict with **p**

```
                { we turn around s counterclockwise }
                V ← this neighbor ;
                the killer of V is x
        endwhile ;
        s ← the third vertex of V
until s = x₁ ;
replace the polygonal chain of Star between x₁ and x₂
        by the edges x₁x and xx₂, associated with W₁ and W₂ ;
the star pointers of W₁ and W₂ reach respectively xx₁ and xx₂ ;
```

6.1.4   Analysis

We assume that p is a random site in S, i.e. p is any of the precedingly inserted sites, with the same probability and independently from the insertion order. More precisely, an event is now one of the $n!$ permutations and one of the n sites. Each event occurs with the same probability $\frac{1}{n.n!}$.

Lemma 6.1 *The expected number of removed nodes is constant.*

Proof. Since p is chosen independently from the insertion order, the expected number of removed nodes is

$$\sum_{T \text{ triangle}} Prob(p \text{ vertex of } T)Prob(T \text{ exists in the Delaunay Tree})$$

$$= \frac{3}{n} \times \text{ expected number of vertices of the Delaunay Tree}$$

$$= O(1)$$

□

Lemma 6.2 *The expected number of unhooked nodes is constant.*

Proof.   In fact the number of unhooked nodes is bounded by the number of edges disappearing in the Delaunay tree, which is :

$$\sum_{T,S \text{ adjacent triangles}} Prob(p \text{ vertex of } T \text{ or } S)$$

$$\times Prob(TS \text{ is an edge of the Delaunay Tree})$$

$$= \frac{4}{n} \times \text{ expected number of edges of the Delaunay Tree}$$

$$= O(1)$$

□

Lemma 6.3 *The expected number of nodes created by the deletion of p is constant.*

Proof. The number of created triangles during the deletion of p is

$$\sum_{T \text{ triangle}} Prob(T \text{ appears during the deletion of p})$$

A triangle T of width j will appear *iff* p is one of the j sites in conflict with T, and p and the 3 vertices of T are introduced before the $j-1$ other sites in conflict with T, and p is not inserted after the 3 vertices of T. So the probability that T appears is :

$$\frac{j}{n} 3 \frac{3!(j-1)!}{(j+3)!} = \frac{3}{n} \frac{3!j!}{(j+3)!}$$

Hence by virtue of the proof of Lemma 4.1, the expected number of triangles created by the deletion of p is :

$$\frac{3}{n} \sum_{T} Prob(T \text{ appears during the insertion phase}) = O(1)$$

$\square$

Lemma 6.4 *The expected cost of a deletion is $O(\log \log n)$.*

Proof.   The expected number of triangles killed by p is constant using Lemma 6.1 (which also implies that the initialization of **Star** is achieved in constant time), and the traversal that is done during the Search step visits a constant number of nodes by Lemma 6.2. For each node, we must locate the creator of the node in **Reinsert**, which can be done in $O(\log \log n)$ worst case deterministic time, by using a bounded ordered dictionary [vEBKZ77]. The universe for this dictionary is the insertion age of the points, or in other words the number of the sites. The required finiteness of the universe can be circumvented using standard dynamization techniques, see for example [Ove83, section5.2].

To preserve the simplicity of the auxiliary data structures we can use a simple balanced binary search tree [AHU83]. In this way we achieve a complexity of $O(\log n)$ time. During the reinsertion phase **Star** can be updated in time proportional to the number of removed nodes.

The total cost of the work on unhooked triangles is constant, since we only have to reach the neighbor of the parent of each of them, and by Lemma 6.2.

For the triangles deleted by the reinsertion of **x**, the cost is linear in the number of triangles in conflict with both p and **x**, which is linear in the number of triangles created by the reinsertion of **x**. By Lemma 6.3, this expected cost is thus constant.

The expected whole cost is then less than $O(\log \log n)$. $\square$

It is important to notice that the randomized hypothesis is preserved by a deletion. Namely, consider now the permutation σ of $S \setminus \{p\}$ obtained by removing p from the insertion order ; there are n permutations of S which give the same σ, so the probability that σ occurs is $n \times \frac{1}{n!}$ and σ is really a random permutation of $S \setminus \{p\}$. The randomization of the $n - 1$ currently present sites is actual and the deletion of p does not affect the analysis of further insertions or deletions. Thus Lemmas 4.1, 4.2 and 6.4 yield the following theorem :

Theorem 6.5 *The Delaunay triangulation (or the Voronoi diagram) of a set S of n sites in the plane can be dynamically maintained in $O(\log n)$ expected time to insert or locate a point and $O(\log \log n)$ expected time to delete a point. This result holds provided that, at any time, the order of insertion on the sites remaining in S may be each order with the same probability, and when a site is deleted, it may be any site with the same probability.*

It is possible to avoid the hypothesis that the random deleted site and the random insertion permutation are independent. It is clear that the deletion of the first inserted site is more expensive that the deletion of the last one, but we show in the sequel that even the deletion of the first inserted site can be done with a good complexity. For this other kind of analysis of deletions, the probability space is only the set of permutations for the insertion, each being equally likely to occur.

Lemma 6.6 *The expected number of removed, unhooked and created nodes during the deletion of the first inserted site is $O(\log n)$.*

> *Proof.* We here only give the proof for the removed nodes, the other quantities can be obtained in the same way. A triangle T of width j exists in the Delaunay tree and is removed during the deletion of the first inserted site if the first site is a vertex of T and if the two other vertices of T are inserted before the j sites in conflict with T. This happens with probability $\frac{3}{n} \frac{2!j!}{(j+2)!}$. By summing for all triangles T, the number of nodes removed in the Delaunay Tree is :
>
> $$\sum_{j=0}^{n-3} |\mathcal{F}_j(S)| \frac{3}{n} \frac{2!j!}{(2+j)!} = O(\log n)$$
>
> □

The same result holds for the kth site : if we do not consider the first site but the kth site, the probability that a triangle is removed during the deletion of the kth site is clearly less than $\frac{3}{n} \frac{2!j!}{(j+2)!}$.

Therefore, the proof of Lemma 6.4 can be modified in a straightforward manner to obtain the following result :

Theorem 6.7 *The expected cost of deleting the kth site is $O(\log n \log \log n)$.*

Thus, for any deletion sequence, the whole set of sites can be deleted in expected time $O(n \log n \log \log n)$, where the expectation is only on the insertion sequence.

6.1.5 d-dimensional case

When the dimension is greater than 2, the general scheme of the algorithm is the same. The difference lies in the fact that the simplices of A form now a polytope of dimension d, that is star-shaped from p. A can still be represented by its boundary. In order to be able to build all the necessary information concerning adjacency on the boundary of A, we must maintain the whole incidence graph of the Delaunay triangulation. The complexity does not increase, though the constants are of course higher.

Though the details are more involved than in the planar case, the analysis extends without any problem, providing an expected cost of $O\left(n^{\lfloor\frac{d+1}{2}\rfloor-1}\log\log n\right)$ for the deletion of a given site.

6.1.6 Practical results in the planar case

The algorithm described here has been effectively coded. This section presents practical results. The sites are first inserted in a random order, and afterwards they are all deleted in another random order. Figures 6.9, 6.10 and 6.11 show the size of the Delaunay Tree in bold line, the size of the Delaunay triangulation in dashed line, and in thin line, a measure of the complexity of the operation. For insertions, it is the number of visited nodes, for deletions it is the number of unhooked triangles plus the numbers of triangles created during this deletion plus the cost of each location in structure **Reinsert**. The cost of deleting a site has a higher variance than the cost of inserting a site ; it may be important if the site had been inserted at the beginning of the construction, but this happens with a low probability.

Even in the case of sites lying on a parabola, which gives a bad behaviour for most of the existing algorithms, our algorithm has a good behaviour in practice.

The Delaunay triangulation of 15000 random sites in a square has been computed in 35 seconds on a Sun 4/75 and the deletion phase has been computed in 50 seconds.

6.2 Removing a segment from an arrangement

This section is roughly similar to the preceding one. We need however to develop some different details (see also [DTY92]).

We assume that the I-DAG has already been constructed as in Section 4.3.2. We need to introduce an auxiliary kind of entity, consisting of the *corners* of trapezoids. When we write that a trapezoid T has at most 4 corners, we now mean that the node of the I-DAG corresponding to T has pointers to those corners. This corners are crucial in the removing process, as will be seen in the sequel. It is not difficult to maintain them, during the insertion step : in Procedure Insert (Figure 4.10) we only have to deduce the corners of the children of T from the corners of T, and to create new corners.

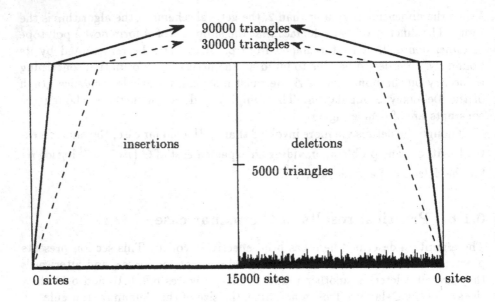

Figure 6.9 : Statistics on 15000 random sites in a square

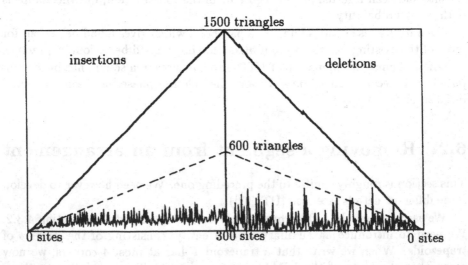

Figure 6.10 : Statistics on 300 random sites on an ellipse

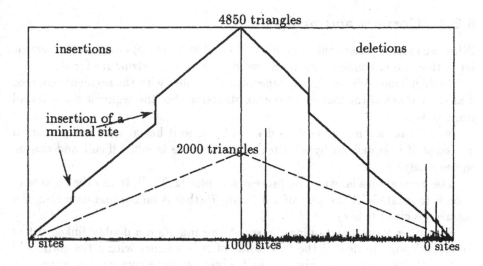

Figure 6.11 : Statistics on 1000 random sites on a parabola

Let us now detail the removing of a segment **s**. As for Delaunay triangulations, we distinguish the unhooked and the removed nodes.

The first part of the removing of **s** will consist in finding all removed nodes. Then we must reinsert the creators of these nodes, while maintaining at each step the set A of trapezoids in conflict with **s**. The unhooked nodes will be processed while processing the removed nodes : the influence range of a trapezoid is included in the influence ranges of its children. So, a removed node cannot have only unhooked nodes. Therefore, an unhooked node must have a removed sibling, and it will be processed in the same time.

The initialization of A is the same as in the preceding section.

6.2.1   The Search step

We do not really look for whole segments to be reinserted. We only look for all parts of them to be reinserted, without trying to connect them. More precisely, what we need to reinsert segments is a list of triplets (x, T, F) consisting of a segment to be reinserted, a node T to be removed, and one parent F of T also to be removed. The reinsertion of such a triplet takes in account only the part of **x** on the boundary of $T \cap F$ (which can be reduced to an extremity of **x**).

We obtain this list by traversing the Influence Graph, starting from the trapezoids in A. Each time we find a node F to be removed, we look at its children, and for each child T defined by **s** and created by the insertion of a segment **x**, we add (x, T, F) to the list.

Furthermore, the segments need to be reinserted in the same order as they had been inserted first, so we need to sort these triplets (the ordering is along **x** only). This can be done in $O(\log \log n)$ time worst case deterministic time as in the preceding section.

6.2.2 Corners and bridges

When we reinsert a segment, we must first of all find the trapezoids in the current set A that are in conflict with it. So we need a location structure for A.

By definition, A is the set of trapezoids in conflict with the segment removed s at some stage of the history of the construction. So, the segment s crosses all trapezoids in A.

Notice that a removed node is defined by s, so it has at least one corner on s (except if it is defined by a vertex of s, this case is not difficult and can be solved easily).

The general idea is, when we process a triplet $(\mathbf{x}, T, F)$, to use the corners of T or F to locate in A the part of $\mathbf{x}$ defining T (that is an approximate idea, the details are given below).

So, rather than maintaining the set A, we maintain a doubly linked list of *bridges* allowing to deduce the trapezoids of A in conflict with $\mathbf{x}$ the corners of T, F or theirs suitable neighbors. The bridges are some corners that appear on s. These bridges are ordered by the order in which they appear on s. More precisely, a corner of a trapezoid is a bridge, if and only if it is on the common boundary of two trapezoids of A, and we store both of them in it (see Figure 6.12). This list of bridges is the analogous of the structure **Star** that we used for the Delaunay triangulation.

In order to initialize the list of bridges, we traverse the set A consisting of all nodes killed by s, by using neighborhood relations. For each node T in A, and each of its children, the corners c of this child that lie on s now appear as bridges on s. T is one of the two trapezoids in A to be stored in c.

6.2.3 Reinsertion of a triplet $(\mathbf{x}, T, F)$

We now describe how a triplet $(\mathbf{x}, T, F)$ is reinserted. These such triplets are processed in the order used for the first insertion of segments, the order for triplets having the same $\mathbf{x}$ is indifferent.

Let us recall that in the triplet $(\mathbf{x}, T, F)$, $\mathbf{x}$ is a segment, T is a removed node created by $\mathbf{x}$ and F is a removed parent of T. At the current stage of the reinsertion, the creator of F is either s or has already been reinserted (because this creator is before $\mathbf{x}$ in the insertion order).

The removed node T has necessarily at least one corner c on the segment s that we delete, we first look for a bridge b *close* to this corner, which means that there is no other bridge on s between b and c. Such a bridge b points to a trapezoid in A in conflict with $\mathbf{x}$. The main problem is to find b using suitable parent and neighbor pointers. To find b we have to look for trapezoid adjacent to s in the old trapezoidal map, trapezoids above and below give two candidate bridges (b_1 and b_2) and b is the closest bridge to c among b_1 and b_2.

Let us assume without loss of generality that c is the down left corner of T. There are several cases to examine.

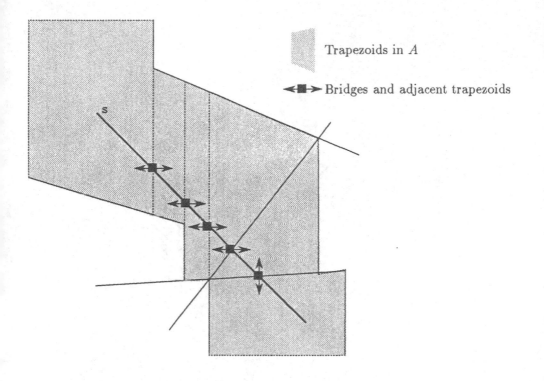

Figure 6.12 : The bridges along s

(1) If the down left corner of F is c then c is already a bridge, so $c = b$ (see Figure 6.13).

(2) If c is not the intersection $s \cap x$, then s is necessarily the down side of T (otherwise, we are in case 1). b_1 is the down left corner of F. b_2 is the left up corner of the down neighbor at creation N of T (see Figure 6.14). b is the closest bridge to c among b_1 and b_2.

(3) If $c = p \cap x$, as in the preceding case b_1 is the down left corner of F. Then, we determine the trapezoid N adjacent to c below s. The down side of T can be s (trapezoid T' in Figure 6.15) or can be x (trapezoid T in Figure 6.15). In the first case N is the down neighbor at creation of T and in the second case N is the down neighbor of the down neighbor at creation of T. N cannot be used directly to determine b_2 since the corners the creator of N is also x and it is not possible to ensure that the corner of N is already a bridge. So we use for b_2 the left up corner of N', parent of N covering the corner $c = x \cap p$ of N. b is the closest bridge to c among b_1 and b_2.

From b we have a direct access to the two trapezoids of A covering b, let S be the one intersecting T. This trapezoid is in conflict with x, so we update the graph in the following way : first x is the (new) killer of S, then x splits S in

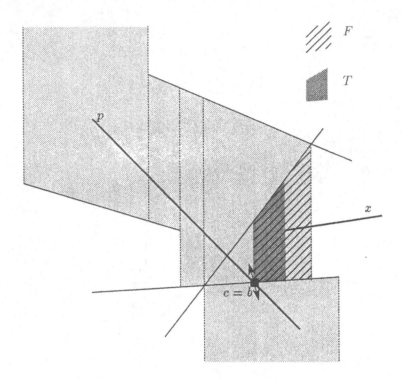

Figure 6.13 : Bridge b is c

pieces, these pieces are children of S and some of them that are in conflict with
s must be created now and added to A. The children of S which are not in
conflict with **s** already exist in the graph and have not to be created, but they
are unhooked and must be hung up to S ; these nodes can be found among the
children of F. Figure 6.16 describes these operations in the case of Figure 6.13.
We must also update the list of bridges, the neighborhood relations and possibly
merge some newly created nodes.

The following pseudo-code procedure summarizes the reinsertion of a triplet.

Reinsert($\mathbf{x}, T, F$)

{wlog, the corner c of T lying on **s** is its down left corner}
{Looking for the closest bridge b to c :}
    if c is already a bridge then $b = c$
    else b_1 = left down corner of F ;
        if $c \neq x \cap p$ then
            b_2 = left up corner of the down neighbor at creation of T
        else {the down neighbor of T is a sibling of T, so its corners are not bridges ⟩
            {in this case we have intersecting points,
               so the stored parent is not any one of the parents (see Section 4.3.2)}

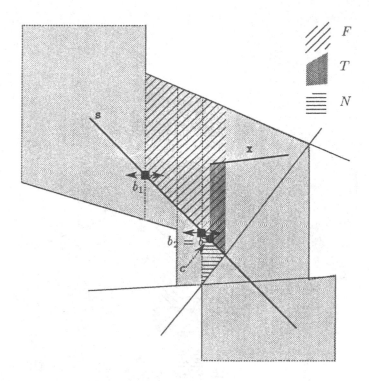

Figure 6.14 : Bridge b is the closest to c among b_1 and b_2

$b_2 =$ left up corner of the parent of the down neighbor of T
    endif ;
    $b =$ closest bridge to c between b_1 and b_2
endif ;

from b, determine the trapezoid S in A which intersects T,
    it is in conflict with x ;
x is the killer of S ;
split S with x ;
for each child R of S
    if R is in conflict with s {R must be added to A}
        if the corner c of T becomes a new bridge (i.e. $b \neq c$)
            create this bridge with R as one of its two trapezoids
            and find its other trapezoid, which is another child of S
            and find its other trapezoid, which is another child of S
            insert c in the doubly linked list of bridges as a successor of b
        else {$b = c$}
            update b : R is one of its two trapezoids
        endif ;

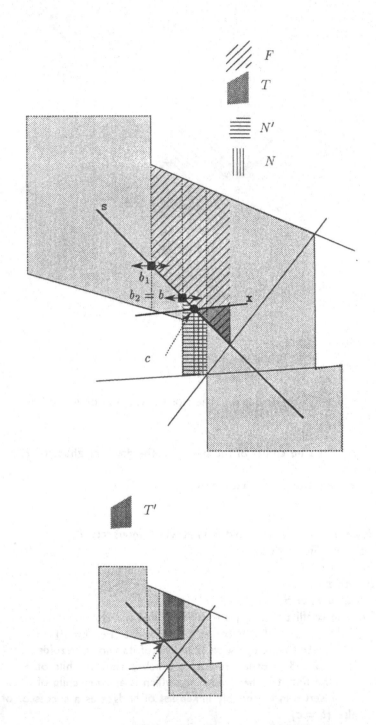

Figure 6.15 : Bridge b is the closest to c among b_1 and b_2

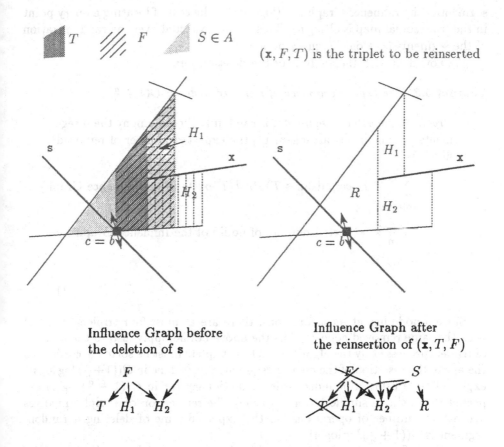

$(\mathbf{x}, F, T)$ is the triplet to be reinserted

Influence Graph before the deletion of **s**

Influence Graph after the reinsertion of $(\mathbf{x}, T, F)$

Figure 6.16 : Computing the new trapezoid R and hanging up the unhooked nodes H_1 and H_2

       else {R must be already existing as an unhooked node}
           find R among the children of F and hang it up to S
       endif ;
       update all other fields in R, and neighbor relations, as for a usual insertion ;
       look for a possible merge with neighbors as for an insertion
   endfor.

6.3 Analysis

The extra work to analyze the deletion of a segment is extremely simple. n is the number of segments currently present in the structure, a is the number of intersection points between these n line segments. Let us first recall the result for insertion proved in Section 4.3.2 : The expected size of the Influence Graph is $O(n+a)$, the expected number of visited nodes during the insertion of the last segment in the Influence Graph is $O(\log n + \frac{a}{n})$, the cost of locating a query point in the trapezoidal map is $O(\log n)$. This cost is expected over the randomization of the segments (not on the query point).

We now deal with the analysis of the deletion phase.

Lemma 6.8 *The expected number of removed nodes is* $O(1 + \frac{a}{n})$.

> *Proof.* If the deleted segment s is randomly chosen among the n segments present in the arrangement, the expected number of removed nodes is
>
> $$\sum_{T \text{ trapezoid}} Prob(p \text{ defines } T) \, Prob(T \text{ exists in the Influence Graph})$$
>
> $$\leq \quad \frac{4}{n} \times \text{ expected number of nodes of the Influence Graph}$$
>
> $$= \quad O(1 + \frac{a}{n})$$
>
> $\square$

Since a node has at most four sons, there are at most four triplets $(\mathbf{x}, T, F)$ with the same removed node F, thus the above bound apply also to the number of triplets processed by the algorithm. These triplets must be sorted according to the age of the insertion of the creator segment, this is done in $O((1 + \frac{a}{n}) \log \log n)$ expected time using a bounded ordered dictionary (or in $O((1 + \frac{a}{n}) \log n)$ expected time using a simpler priority queue). The reinsertion of a triplet involves a constant number of operations, so the expected time of deleting a random segment is $O((1 + \frac{a}{n}) \log \log n)$.

As the Influence Graph is restored as if s was never inserted, the results stated for the insertion step are still valid after a segment has been removed, and we obtain the main result of this section :

Theorem 6.9 *An arrangement of line segments in the plane can be maintained dynamically in* $O(\log n + \frac{a}{n})$ *time for an insertion and* $O((1 + \frac{a}{n}) \log \log n)$ *time for a deletion. The space complexity is* $O(n + a)$. *All complexities are expected with respect to the randomized insertion of the line segments. Here n denotes the current number of segments present in the arrangement, and a denotes the current complexity of the arrangement.*

The expected cost of the location of any point in the arrangement is $O(\log n)$, *where the expectation is over the randomization of the order of insertion of the line segments present in the arrangement, these queries can be done persistently with respect to the insertions.*

Conclusion

The common points between these two examples seem to reside only in the general idea of reconstructing the past. However, we can also see similarities in the analysis.

Other approaches appeared recently in the literature, they will be rapidly presented in Chapter 7.

Chapter 7

Parallel work

We rapidly present in this chapter some papers that have mostly been published recently or even not published yet, in order to replace our work amidst the community.

Only general ideas are given, without any detail.

7.1 Accelerated static algorithms

The union between conflict graph and influence graph permits to accelerate randomized algorithms, under some hypotheses.

R. Seidel was the first one to use such a combination, to compute the trapezoidal map induced by a simple polygonal chain of size n [Sei91b]. He deduces an algorithm running in $O(n \log^* n)$ expected time (more generally, $O(n \log^* n + k \log n)$ when the segments form a planar graph with k connected components).

O. Devillers [Dev92] generalized this result to many cases, when similar informations are given (some traversal order of the segments) about the data : he deals for example with the computation of the skeleton of a simple polygon, or the Delaunay triangulation of a set of points knowing the minimum spanning tree.

Let us recall that for $n > 0$, $\log^* n$ is the largest integer l such that $\log^{(l)} n \leq 1$, where $x \mapsto \log^{(l)} x$ is the application obtained by l iterations of function $x \mapsto \log x$. It is an extremely slowly increasing function ($\log^* n < 5$ for $n < 2^{65536}$).

The idea of the algorithm is the following : Let $s_1, s_2, \ldots, s_n$ be a random permutation among the $n!$ permutations of set S of line segments. The process is initialized by computing the trapezoidal map $\mathcal{T}_1$ induced by the first segment s_1. h steps are then carried out, consisting, at each step $i \geq 2$, in :

- inserting $N_i - N_{i-1}$ segments in the trapezoidal map $\mathcal{T}_{i-1}$, and constructing and influence graph. A trapezoidal map $\mathcal{T}_i$ is obtained.

- building the conflict graph between the $n - N_i$ uninserted segments and the trapezoids in $\mathcal{T}_i$, by using the traversal order given by the known informations (for example, the segments form a simple polygon). This is achieved by traversing $\mathcal{T}_i$ using the neighborhood relations.

When a new segment is inserted, at the next step $i + 1$, the conflict graph of trapezoids in $\mathcal{T}_i$ is used to find the conflicts with this map, then the conflicts with more recent trapezoids are obtained by traversing only the part of the influence graph that is just being constructed at that step.

This allows us to use point *(3)* of Theorem 4.9. A judicious choice of parameters ($h = \log^* n$ and $N_i = \left\lfloor \frac{n}{\log^{(i)} n} \right\rfloor$) gives the announced complexity.

7.2   Semi-dynamic algorithms

Computing one cell in an arrangement of line segments

In [CEGSS91], B. Chazelle, H. Edelsbrunner, L. Guibas, M. Sharir and J. Snoeyink compute a particular cell in an arrangement of n line segments. The
randomized expected complexity of the algorithm is $O(n\alpha(n)\log n)$, while the
best deterministic algorithm known was running in $O(n\alpha(n)\log^2 n)$ worst-case
time. $\alpha(n)$ is the pseudo-inverse of Ackermann function ; α grows to infinite, but
excessively slowly, and remains less than 3 for any "reasonable" value (smaller
than $2^{2^{2^{.^{.^{.^2}}}}} \nearrow {}^{2^{65536}}$) of n. Let us recall that the worst-case size of a cell is $\Omega(n\alpha(n))$.

The algorithm is semi-dynamic, it works by maintaining the history of the
construction of the trapezoidal map induced by the segments, in a graph similar
to the I-DAG. To detect whether a new segment intersects the boundary of the
desired cell, the connected components of its boundary are stored in a classical *union-find* structure, allowing to update the trapezoidal map when a new
segment is inserted.

7.3   Dynamic algorithms

Several authors recently worked on the possibility of performing deletions in
randomized structures.

7.3.1   With storage of the history

General approaches

O. Schwarzkopf [Sch91] builds a larger history of the structure, keeping track of
deletions as well as insertions. In this way, he obtains a structure whose memory
cost is bigger. So he must reconstruct it, whenever its size is too big, by only
keeping objects that have been inserted and not removed at the current step.

In the particular case of Voronoi diagram in the plane, each insertion is done
in $O(\log n)$ expected time, and locating a points takes $O(\log^2 n)$ time with high
probability.

The construction of the trapezoidal map of a set of n non-crossing line segments can also be treated in $O(\log^2 n)$ expected time for an insertion.

F. Aurenhammer and O. Schwarzkopf use a similar method for higher orders
Voronoi diagrams [AS92].

In a new paper, K. Mulmuley takes the same scheme as O. Schwarzkopf
[Mul91c]. The difference is that the structure is not based upon the real history of
the construction, but on another, imaginary, sequence of insertions and deletions.
When this history is too large, he reconstructs another one. Another, more
powerful, alternative, is to rebalance the structure, by doing operations similar
to rotations in binary search trees, still by manipulating this imaginary past.

Convex hulls

K.L. Clarkson, K. Mehlhorn and R. Seidel [CMS92] solve the problem for convex hulls in any dimension. They use the same general principle as we use : to reconstruct a new past. But the history of the construction is maintained in the convex hull itself : more precisely, the authors maintain a triangulation of the convex hull, in which they can locate the new point to be inserted.

If this point is interior to the current convex hull, that is if it belongs to a simplex of its triangulation, we store this information. If it is exterior, we construct the simplices formed by this point and the visible facets of the convex hull, and we add them to the hull.

When a vertex of the hull is removed, the triangulation also gives the interior points that might appear as new vertices : they are points interior to the simplices incident to this vertex. These points will be reinserted. When a vertex of the triangulation is removed, some of the points interior to the simplices incident to this vertex must also been reinserted, in order to reconstruct the history.

A randomized analysis gives a $O(\log n)$ expected complexity in dimension ≤ 3, and $O\left(n^{\lfloor \frac{d}{2} \rfloor - 1}\right)$ in higher dimensions, for each operation.

7.3.2 Without storing the history

K. Mulmuley [Mul91b, MS91] uses a radically different approach, and successfully avoids the storage of the history, recapturing the idea of [AS89].

He constructs a hierarchical structure : on level 1, all objects are present. Then for any $i \leq 2$, the objects present on level i are selected by tossing a coin for each object of level $i - 1$. The geometric structure is computed for each level.

Between two levels, the conflicts of a region determined by objects of level i, with the objects of level $i - 1$ not belonging to level i, are stored in a conflict graph. In addition, a *Descent* structure gives the location of an object of level i, knowing its location in level $i - 1$.

When a new object is inserted, the coin is tossed, the object is inserted in each level, up to a negative answer. The structures between levels must be updated. Removing an object is now only the exact opposite of an insertion. The only difference is that an insertion must firstly look for conflicts, which increases the complexity.

This idea is applied to the dynamic construction of the Delaunay triangulation in the plane. The expected complexities are $O(\log n)$ for an insertion, $O(1)$ for a deletion, and $O(\log^2 n)$ for a location. For the case of arrangements of line segments in the plane, an insertion runs in $O(\log a)$ expected time, a deletion in $O(1)$ expected time, and a location in $O(\log a)$ with high probability, where a is the size of the trapezoidal map. The analyses use random sampling.

7.4   Strategies

Several authors recently studied problems of a different kind. If the running time t_A of an algorithm A is given by its expectation, Markov's inequality gives the probability that the running time be superior to a given value. It is the best known bound when no additional information on the probabilistic distribution of t_A is known.

Let us now assume that the following experiment is performed : It runs A for t_1 time units, then, if A does not stop before t_1, it runs A again for t_2 time units, and so on. Thus a new algorithm is obtained, and we look for the optimal sequence $t_1, t_2, \ldots$. H. Alt, L. Guibas, K. Mehlhorn, R. Karp and A. Widgerson prove, among other results, that there is always an optimal strategy, if the expectation of t_A is known [AGMKW91].

K. Mehlhorn, M. Sharir and E. Welzl study a similar problem, but they compute the space complexity, and obtain very tight tail estimates [MSW92].

Conclusion

We presented here different trends for data structures and algorithms, in the field of dynamic randomized algorithms. As publications in this field are rushing, this presentation deliberately stop at what I have heard of, up to this fatidical day of october 14th, 1991 ! I have no doubt that the field will still inspire researchers in the future.

Conclusion

Our paper about the Delaunay Tree [BT86] claimed an expected complexity of $O(\log n)$ per insertion, in the planar case, as well as in 3D when the size of the Delaunay triangulation is linear. The main problem in ouR proofs was the lack of a clear assumption about the insertion order of the points. This kind of assumption was not usual at that time. It was only mentioned in experimental statements : as soon as the points were inserted in random order, even very strongly degenerated distributions were treated with the same efficiency as homogeneous distributions. Hindsight allows us to see these proofs more clearly, and to realize that their central arguments were roughly the same as the ones now common in randomized algorithms.

Let us emphasize these arguments : n points are assumed to be inserted according to a random permutation ω among the $n!$ possible ones. We can restrict ourselves to the planar case, since the same proofs still hold in 3D when the size of the triangulation is linear. Euler's formula gives a way to bound by 6 the average degree of a point in the Delaunay triangulation, by dividing the number of edges by the number of points. It was deduced that the expected number of triangles created by the insertion of the kth point o was at most 6. This very argument, which uses "backwards" analysis technique, can be found in Lemma 4.6 : o is any of the k points present after its own insertion, since ω is any permutation. Thus, it creates any of the triangles present at step k with probability $\frac{3}{k}$.

Another crucial point consisted of saying that, when the lth point site o was inserted, the number of triangles present at step k $(k < l)$ and in conflict with o was the same as it would have been if o had been inserted as the $(k + 1)^{th}$ site. The idea behind Lemma 4.5 can be recognized, it is only an averaging on all possible permutations.

However, we had to wait for [Dev92] and the "backwards" analysis introduced by R. Seidel, to settle our proofs of 1986 rigorously. Meanwhile, K.L. Clarkson presented the new technique of random sampling, and we followed him, as did almost the whole international community. It might have been possible for us to reach directly a backwards analysis.

Our work proved that the approach followed in the design of the Delaunay Tree —storing the history of the construction done by the algorithm— was good, since we have been able to generalize this structure in two different ways by applying K.L. Clarkson's formalism : the Influence Graph allows us to compute various geometric structures, provided that they can be defined as sets of regions without conflicts ; the k-Delaunay Tree extends the possibilities to the case of regions having at most k conflicts. The Influence Graph has also been shown to be a fully dynamic structure.

One quality of randomized incremental algorithms is their simplicity, copiously proved in recent literature, of which our algorithms are only an additional illustration.

The comparison of algorithms using the Influence Graph with similar work

—many using the same approach based on the history (I cannot refrain myself from citing here Otfried Schwarzkopf [Sch92] who qualifies our idea [BT86] as "belonging to the folklore of the field")— shows the theoretical efficiency of this structure. The Delaunay Tree was the first data structure for designing efficient semi-dynamic geometric algorithms, it was also one of the very first dynamic structures. The implementation of these algorithms also showed their practical efficiency.

In spite of the efficiency of randomized algorithms, research on M deterministic algorithms has not stopped. Since 1988, B. Chazelle et J. Friedman, for example, have taken an interest in "derandomizing" algorithms based on random sampling and divide-and-conquer scheme [CF90]. much of this research is only of theoretical interest, because optimal deterministic algorithms are often very complicated and difficult to implement. It is however useful, because the discovery of any deterministic algorithm, even a non practical one, proves the existence of a solution, and opens the way towards simpler solutions.

In [BT86], the question was raised to search for an optimal order for inserting sites in the Delaunay Tree, so that a maximal cost for any location could be guaranteed; the randomized analysis only gives an expected cost. The algorithm would then loose its "on-line" property, since the preliminary knowledge of the whole set of data would be necessary to find this order. By asking the same question, B. Chazelle has just found an optimal algorithm to compute convex hulls in any dimension [Cha91], which until then was an open problem, though convex hull was one of the most studied structures in geometry.

The most interesting algorithms in practice are probably output-sensitive algorithms, more than optimal algorithms, that are only worst-case optimal.

In fact, it has been shown that randomized incremental algorithms are sensitive, not to the final output size, but to the size of all intermediate results. When the size of the studied structures grows with the number of input data, this is sufficient. This holds for example for the construction of the trapezoidal map of an arrangement of line segments in the plane.

However for the problem of hidden line or hidden surface removal, for example, this approach cannot give good results : let us assume that a big object hides a scene of quadratic complexity. This object will be inserted, with probability $\frac{1}{2}$, among one of the $\frac{n}{2}$ last objects. Numerous intermediate results have thus a quadratic size, while the final output has constant size. The design of output-sensitive algorithms is crucial in this field. Let us for example mention the paper by M. de Berg et M. Overmars [dBO91] who gave one of the first simple solutions to this problem.

There are several deterministic output-sensitive algorithms, for the geometric structures studied in this thesis. The Delaunay triangulation, from dimension 3, is another example raising the same problem : even if the size of the final triangulation is linear, some intermediate triangulation may have a quadratic size. In [Boi88, BCDT91], we obtain an optimal algorithm $O(n \log n + t)$ (n is the number of point sites, and t the number of tetrahedra in the triangulation), for the restricted case when the points are assumed to lie on two planes. The algorithm by R. Seidel [Sei86] for the computation of convex hulls in any dimension runs in $O(n^2 + f \log n)$, where f is the numbers of faces of the hull.

Too few results exist in this field, and it can be supposed that this subject will be much studied in future.

Bibliography

[AES85]   D. Avis, H. ElGindy, and R. Seidel. Simple on-line algorithms
            for convex polygons. In G.T. Toussaint, editor, *Computational
            Geometry*, pages 23–42. North Holland, 1985.

[AGMKW91]   H. Alt, L. Guibas, K. Mehlhorn, R. Karp, and A. Widgerson.
            A method for obtaining randomized algorithms with small tail
            probabilities, 1991. To appear.

[AGSS89]   A. Aggarwal, L.J. Guibas, J. Saxe, and P.W. Shor. A linear time
            algorithm for computing the Voronoi diagram of a convex polygon.
            Discrete and Computational Geometry, 4:591–604, 1989.

[AHU83]   A. Aho, J. Hopcroft, and J. Ullman. *The design and ananlysis of
            computer algorithms*. Computer science and information process-
            ing. Addison Wesley, 1983.

[AS89]   C. Aragon and R. Seidel. Randomized search trees. In *IEEE
            Symposium on Foundations of Computer Science*, pages 540–545,
            1989.

[AS92]   F. Aurenhammer and O. Schwarzkopf. A simple on-line random-
            ized incremental algorithm for computing higher order Voronoi
            diagrams. *International Journal of Computational Geometry and
            Applications*, 2(4):363–382, 1992.

[Aur91]   F. Aurenhammer. Voronoi diagrams — a survey of a fundamental
            geometric data structure. *ACM Computing Surveys*, 23(3):345–
            405, September 1991.

[BCDT91]   J-D. Boissonnat, A. Cérézo, O. Devillers, and M. Teillaud. Output
            sensitive construction of 3D Delaunay triangulation of constrained
            sets of points. In *Third Canadian Conference on Computational
            Geometry in Vancouver*, pages 110–113, August 1991. Full paper
            available as Technical Report INRIA 1415. To appear in IJCGA.

[BD92]   J-D. Boissonnat and K. Dobrindt. Randomized construction of
            the upper envelope of surface patches in three dimensions. In

Fourth Canadian Conference on Computational Geometry, pages 311–315, 1992. Full paper available as Technical Report INRIA 1878.

[BDSTY92] J-D. Boissonnat, O. Devillers, R. Schott, M. Teillaud, and M. Yvinec. Applications of random sampling to on-line algorithms in computational geometry. *Discrete and Computational Geometry*, 8:51–71, 1992.

[BDT93] J-D. Boissonnat, O. Devillers, and M. Teillaud. A semi-dynamic construction of higher order Voronoi diagrams and its randomized analysis. *Algorithmica*, 9:329–356, 1993.

[Boi88] J-D. Boissonnat. Shape reconstruction from planar cross-sections. *Computer Vision, Graphics, and Image Processing*, 44:1–29, 1988.

[Bow81] A. Bowyer. Computing Dirichlet tesselations. *The Computer Journal*, 24(2), 1981.

[Bro79] K.Q. Brown. Voronoi diagrams from convex hulls. *Information Processing Letters*, 9:223–228, 1979.

[BT86] J-D. Boissonnat and M. Teillaud. A hierarchical representation of objects: The Delaunay Tree. In *Second ACM Symposium on Computational Geometry in Yorktown Heights*, pages 260–268, June 1986.

[BT93] J-D. Boissonnat and M. Teillaud. On the randomized construction of the Delaunay tree. *Theoretical Computer Science*, 112:339–354, 1993.

[CE85] B. Chazelle and H. Edelsbrunner. An improved algorithm for constructing k^{th}-order Voronoi diagrams. In *First ACM Symposium on Computational Geometry in Baltimore*, pages 228–234, June 1985.

[CE88] B. Chazelle and H. Edelsbrunner. An optimal algorithm for intersecting line segments in the plane. In *IEEE Symposium on Foundations of Computer Science*, pages 590–600, 1988.

[CEGSW90] K.L. Clarkson, H. Edelsbrunner, L.J. Guibas, M. Sharir, and E. Welzl. Combinatorial complexity bounds for arrangements of curves and surfaces. *Discrete and Computational Geometry*, 5:99–160, 1990.

[CEGSS91] B. Chazelle, H. Edelsbrunner, L.J. Guibas, M. Sharir, and J. Snoeyink. Computing a face in an arrangement of line segments. In *ACM-SIAM Symposium on Discrete Algorithms*, pages 441–448, 1991.

[CF90]   B. Chazelle and J. Friedman. A deterministic view of random sampling and its use in geometry. *Combinatorica*, 10:229–249, 1990.

[Cha91]   B. Chazelle. An optimal convex hull algorithm for point sets in any fixed dimension. Technical Report CS-TR-336-91, Computer Science Department, Princeton University, (USA), June 1991.

[Cla85]   K.L. Clarkson. A probabilistic algorithm for the post office problem. In *17th Annual SIGACT Symposium*, pages 75–184, 1985.

[Cla87]   K.L. Clarkson. New applications of random sampling in computational geometry. *Discrete and Computational Geometry*, 2:195–222, 1987.

[CMS92]   K.L. Clarkson, K. Mehlhorn, and R. Seidel. Four results on randomized incremental constructions. In *LNCS 577 (STACS 92)*, pages 463–474, Springer-Verlag, 1992.

[CS88]   K.L. Clarkson and P.W. Shor. Algorithms for diametral pairs and convex hulls that are optimal, randomized, and incremental. In *4th ACM Symposium on Computational Geometry in Urbana*, pages 12–19, 1988.

[CS89]   K.L. Clarkson and P.W. Shor. Applications of random sampling in computational geometry, II. *Discrete and Computational Geometry*, 4(5):387–421, 1989.

[dBO91]   M. de Berg and M.H. Overmars. Hidden surface removal for axis-parallel polyhedra. In *IEEE Symposium on Foundations of Computer Science*, pages 252–261, 1991.

[Dev92]   O. Devillers. Randomization yields simple $O(n \log^* n)$ algorithms for difficult $\Omega(n)$ problems. *International Journal of Computational Geometry and Applications*, 2(1):97–111, March 1992.

[DMT92a]   O. Devillers, S. Meiser, and M. Teillaud. Fully dynamic Delaunay triangulation in logarithmic expected time per operation. *Computational Geometry Theory and Applications*, 2(2):55–80, 1992.

[DMT92b]   O. Devillers, S. Meiser, and M. Teillaud. The space of spheres, a geometric tool to unify duality results on Voronoi diagrams. *Fourth Canadian Conference on Computational Geometry*, pages 263–268, 1992. Full paper available as Technical Report INRIA 1620.

[DTY92]   O. Devillers, M. Teillaud, and M. Yvinec. Dynamic location in an arrangement of line segments in the plane. *Algorithms Review*, 2(3), March 1992. Newsletter of ALCOM project.

[Dwy91]   R.A. Dwyer. Higher-dimensional Voronoi diagrams in linear ex-
             pected time. *Discrete and Computational Geometry*, 6:343–367,
             1991.

[Ede87]   H. Edelsbrunner. *Algorithms on Combinatorial Geometry*. Sprin-
             ger-Verlag, 1987.

[EOS86]   H. Edelsbrunner, J. O'Rourke, and R. Seidel. Constructing ar-
             rangements of lines and hyperplanes with applications. *SIAM
             Journal on Computing*, 15:341–363, 1986.

[ES74]   P. Erdös and J. Spencer. *Probabilistic Methods in Combinatorics*.
             Academic Press, 1974.

[ES86]   H. Edelsbrunner and R. Seidel. Voronoi diagrams and arrange-
             ments. *Discrete and Computational Geometry*, 1:25–44, 1986.

[ESS91]   H. Edelsbrunner, R. Seidel, and M. Sharir. On the Zone Theorem
             for hyperplane arrangements. In *LNCS 555 (New results and new
             trends in Computer Science)*, 108–123, 1991.

[GKS92]   L.J. Guibas, D.E. Knuth, and M. Sharir. Randomized incremental
             construction of Delaunay and Voronoi diagrams. *Algorithmica*,
             7(4):381–413, 1992.

[GS78]   P.J. Green and R. Sibson. Computing Dirichlet tesselations in the
             plane. *The Computer Journal*, 21, 1978.

[Kle80]   V. Klee. On the complexity of d-dimensional Voronoi diagrams.
             Archiv der Mathematik, 34:75–80, 1980.

[Lee82]   D.T. Lee. On *k*-nearest neighbor Voronoi diagrams in the plane.
             IEEE Transactions on Computers, C-31:478–487, 1982.

[Meh84]   K. Mehlhorn. *Data Structures and Algorithms 3: Multidimen-
             sional Searching and Computational Geometry*. Springer-Verlag,
             1984.

[MMO91]   K. Mehlhorn, S. Meiser, and C. Ó'Dúnlaing. On the construc-
             tion of abstract Voronoi diagrams. *Discrete and Computational
             Geometry*, 6:211–224, 1991.

[MS91]   K. Mulmuley and S. Sen. Dynamic point location in arrange-
             ments of hyperplanes. In *7th ACM Symposium on Computational
             Geometry in North Conway*, pages 132–142, 1991.

[MSW92]   K. Mehlhorn, M. Sharir, and E. Welzl. Tail estimates for the space
             complexity of randomized incremental algorithms. In *ACM-SIAM
             Symposium on Discrete Algorithms*, January 1992.

[Mul88]   K. Mulmuley. A fast planar partition algorithm, I. In *IEEE Symposium on Foundations of Computer Science*, pages 580–589, 1988.

[Mul89a]   K. Mulmuley. A fast planar partition algorithm, II. In *5th ACM Symposium on Computational Geometry in Saarbrücken*, pages 33–43, 1989.

[Mul89b]   K. Mulmuley. On obstruction in relation to a fixed viewpoint. In *IEEE Symposium on Foundations of Computer Science*, pages 592–597, 1989.

[Mul91a]   K. Mulmuley. On levels in arrangements and Voronoi diagrams. *Discrete and Computational Geometry*, 6:307–338, 1991.

[Mul91b]   K. Mulmuley. Randomized multidimensional search trees : Dynamic sampling. In *7th ACM Symposium on Computational Geometry in North Conway*, pages 121–131, 1991.

[Mul91c]   K. Mulmuley. Randomized multidimensional search trees: lazy balancing and dynamic shuffling. In *IEEE Symposium on Foundations of Computer Science*, pages 180–196, 1991.

[Ove83]   M.H. Overmars. *The design of dynamic data structures*. LNCS 156. Springer-Verlag, 1983.

[PS85]   F.P. Preparata and M.I. Shamos. *Computational Geometry : an Introduction*. Springer-Verlag, 1985.

[Raj91]   V.T. Rajan. Optimality of the Delaunay triangulation in $\mathbb{R}^d$. In *7th ACM Symposium on Computational Geometry in North Conway*, pages 357–363, June 1991.

[Sch91]   O. Schwarzkopf. Dynamic maintenance of geometric structures made easy. In *IEEE Symposium on Foundations of Computer Science*, pages 197–206, October 1991. Full paper available as Technical Report B 91-05 Universität Berlin.

[Sch92]   O. Schwarzkopf. *Dynamic maintenance of convex polytopes and related structures*. PhD Thesis, Fachbereich Mathematik, Freie Universität Berlin, June 1992.

[Sei81]   R. Seidel. A convex hull algorithm optimal for point sites in even dimensions. Technical Report 14, Departement of Computer Science, University British Columbia, Vancouver, BC, 1981.

[Sei86]   R. Seidel. Constructing higher-dimensional convex hulls at logarithmic cost per face. In *ACM Symposium on Theory of Computing*, pages 404–413, 1986.

[Sei91a]   R. Seidel. Backwards analysis of randomized geometric algo-
             rithms, June 1991. Manuscript. *ALCOM Summerschool on ef-
             ficient algorithms design*, Aarhus, Denmark.

[Sei91b]   R. Seidel. A simple and fast randomized algorithm for computing
             trapezoidal decompositions and for triangulating polygons. *Com-
             putational Geometry Theory and Applications*, 1(1):51–64, 1991.

[Sei91c]   R. Seidel. Small-dimensional linear programming and convex hulls
             made easy. *Discrete and Computational Geometry*, 6:423–434,
             1991.

[SH75]   M.I. Shamos and D. Hoey. Closest-point problems. In *IEEE
             Symposium on Foundations of Computer Science*, pages 151–162,
             October 1975.

[Sha78]   M.I. Shamos. *Computational Geometry*. PhD thesis, Department
             of Computer Science, Yale University, (USA), 1978.

[vEBKZ77]   P. van Emde Boas, R. Kaas, and E. Zijlstra. Design and imple-
             mentation of an efficient priority queue. *Mathematical Systems
             Theory*, 10:99–127, 1977.

[Wel85]   E. Welzl. Constructing the visibility graph for n line segments in
             $O(n^2)$ time. *Information Processing Letters*, 20:167–171, 1985.

[Yap87]   C.K. Yap. An $O(n \log n)$ algorithm for the Voronoi diagram of a
             set of simple curve segments. *Discrete and Computational Geom-
             etry*, 2:365–393, 1987.

Index

Springer-Verlag
and the Environment

We at Springer-Verlag firmly believe that an international science publisher has a special obligation to the environment, and our corporate policies consistently reflect this conviction.

We also expect our business partners – paper mills, printers, packaging manufacturers, etc. – to commit themselves to using environmentally friendly materials and production processes.

The paper in this book is made from low- or no-chlorine pulp and is acid free, in conformance with international standards for paper permanency.

Lecture Notes in Computer Science

For information about Vols. 1–680
please contact your bookseller or Springer-Verlag

Vol. 717: I. Sommerville, M. Paul (Eds.), Software Engineering – ESEC '93. Proceedings, 1993. XII, 516 pages. 1993.

Vol. 718: J. Seberry, Y. Zheng (Eds.), Advances in Cryptology – AUSCRYPT '92. Proceedings, 1992. XIII, 543 pages. 1993.

Vol. 719: D. Chetverikov, W.G. Kropatsch (Eds.), Computer Analysis of Images and Patterns. Proceedings, 1993. XVI, 857 pages. 1993.

Vol. 720: V.Mařík, J. Lažanský, R.R. Wagner (Eds.), Database and Expert Systems Applications. Proceedings, 1993. XV, 768 pages. 1993.

Vol. 721: J. Fitch (Ed.), Design and Implementation of Symbolic Computation Systems. Proceedings, 1992. VIII, 215 pages. 1993.

Vol. 722: A. Miola (Ed.), Design and Implementation of Symbolic Computation Systems. Proceedings, 1993. XII, 384 pages. 1993.

Vol. 723: N. Aussenac, G. Boy, B. Gaines, M. Linster, J.-G. Ganascia, Y. Kodratoff (Eds.), Knowledge Acquisition for Knowledge-Based Systems. Proceedings, 1993. XIII, 446 pages. 1993. (Subseries LNAI).

Vol. 724: P. Cousot, M. Falaschi, G. Filè, A. Rauzy (Eds.), Static Analysis. Proceedings, 1993. IX, 283 pages. 1993.

Vol. 725: A. Schiper (Ed.), Distributed Algorithms. Proceedings, 1993. VIII, 325 pages. 1993.

Vol. 726: T. Lengauer (Ed.), Algorithms – ESA '93. Proceedings, 1993. IX, 419 pages. 1993

Vol. 727: M. Filgueiras, L. Damas (Eds.), Progress in Artificial Intelligence. Proceedings, 1993. X, 362 pages. 1993. (Subseries LNAI).

Vol. 728: P. Torasso (Ed.), Advances in Artificial Intelligence. Proceedings, 1993. XI, 336 pages. 1993. (Subseries LNAI).

Vol. 729: L. Donatiello, R. Nelson (Eds.), Performance Evaluation of Computer and Communication Systems. Proceedings, 1993. VIII, 675 pages. 1993.

Vol. 730: D. B. Lomet (Ed.), Foundations of Data Organization and Algorithms. Proceedings, 1993. XII, 412 pages. 1993.

Vol. 731: A. Schill (Ed.), DCE – The OSF Distributed Computing Environment. Proceedings, 1993. VIII, 285 pages. 1993.

Vol. 732: A. Bode, M. Dal Cin (Eds.), Parallel Computer Architectures. IX, 311 pages. 1993.

Vol. 733: Th. Grechenig, M. Tscheligi (Eds.), Human Computer Interaction. Proceedings, 1993. XIV, 450 pages. 1993.

Vol. 734: J. Volkert (Ed.), Parallel Computation. Proceedings, 1993. VIII, 248 pages. 1993.

Vol. 735: D. Bjørner, M. Broy, I. V. Pottosin (Eds.), Formal Methods in Programming and Their Applications. Proceedings, 1993. IX, 434 pages. 1993.

Vol. 736: R. L. Grossman, A. Nerode, A. P. Ravn, H. Rischel (Eds.), Hybrid Systems. VIII, 474 pages. 1993.

Vol. 737: J. Calmet, J. A. Campbell (Eds.), Artificial Intelligence and Symbolic Mathematical Computing. Proceedings, 1992. VIII, 305 pages. 1993.

Vol. 738: M. Weber, M. Simons, Ch. Lafontaine, The Generic Development Language Deva. XI, 246 pages. 1993.

Vol. 739: H. Imai, R. L. Rivest, T. Matsumoto (Eds.), Advances in Cryptology – ASIACRYPT '91. X, 499 pages. 1993.

Vol. 740: E. F. Brickell (Ed.), Advances in Cryptology – CRYPTO '92. Proceedings, 1992. X, 593 pages. 1993.

Vol. 741: B. Preneel, R. Govaerts, J. Vandewalle (Eds.), Computer Security and Industrial Cryptography. Proceedings, 1991. VIII, 275 pages. 1993.

Vol. 742: S. Nishio, A. Yonezawa (Eds.), Object Technologies for Advanced Software. Proceedings, 1993. X, 543 pages. 1993.

Vol. 743: S. Doshita, K. Furukawa, K. P. Jantke, T. Nishida (Eds.), Algorithmic Learning Theory. Proceedings, 1992. X, 260 pages. 1993. (Subseries LNAI)

Vol. 744: K. P. Jantke, T. Yokomori, S. Kobayashi, E. Tomita (Eds.), Algorithmic Learning Theory. Proceedings, 1993. XI, 423 pages. 1993. (Subseries LNAI)

Vol. 745: V. Roberto (Ed.), Intelligent Perceptual Systems. VIII, 378 pages. 1993. (Subseries LNAI)

Vol. 746: A. S. Tanguiane, Artificial Perception and Music Recognition. XV, 210 pages. 1993. (Subseries LNAI)

Vol. 747: M. Clarke, R. Kruse, S. Moral (Eds.), Symbolic and Quantitative Approaches to Reasoning and Uncertainty. Proceedings, 1993. X, 390 pages. 1993.

Vol. 748: R. H. Halstead Jr., T. Ito (Eds.), Parallel Symbolic Computing: Languages, Systems, and Applications. Proceedings, 1992. X, 419 pages. 1993.

Vol. 749: P. A. Fritzson (Ed.), Automated and Algorithmic Debugging. Proceedings, 1993. VIII, 369 pages. 1993.

Vol. 750: J. L. Diaz-Herrera (Ed.), Software Engineering Education. Proceedings, 1994. XII, 601 pages. 1994.

Vol. 751: B. Jähne, Spatio-Temporal Image Processing. XII, 208 pages. 1993.

Vol. 752: T. W. Finin, C. K. Nicholas, Y. Yesha (Eds.), Information and Knowledge Management. Proceedings, 1992. VII, 142 pages. 1993.

Vol. 753: L. J. Bass, J. Gornostaev, C. Unger (Eds.), Human-Computer Interaction. Proceedings, 1993. X, 388 pages. 1993.

Vol. 755: B. Möller, H. Partsch, S. Schuman (Eds.), Formal Program Development. Proceedings. VII, 371 pages. 1993.

Vol. 756: J. Pieprzyk, B. Sadeghiyan, Design of Hashing Algorithms. XV, 194 pages. 1993.

Vol. 758: M. Teillaud, Towards Dynamic Randomized Algorithms in Computational Geometry. IX, 157 pages. 1993.